Thomas Rawson Birks

First Principles of Moral Science

Thomas Rawson Birks
First Principles of Moral Science
ISBN/EAN: 9783337035747
Printed in Europe, USA, Canada, Australia, Japan
Cover: Foto ©Thomas Meinert / pixelio.de

More available books at **www.hansebooks.com**

FIRST PRINCIPLES

OF

MORAL SCIENCE,

A COURSE OF LECTURES

DELIVERED IN

The University of Cambridge

BY

THOMAS RAWSON BIRKS,

KNIGHTBRIDGE PROFESSOR OF MORAL PHILOSOPHY.

London:
MACMILLAN AND CO.
1873.

PREFACE.

THIS small volume contains the substance of a first course of Lectures, delivered in October and November 1872, in partial fulfilment of the honourable trust a few months earlier confided to me, after the lamented death of Professor Maurice.

The subjects formally assigned to the Knightbridge Professor are Moral Philosophy, Casuistry, and Moral Theology. The first and third of these scarcely need any exposition of their meaning. The second was viewed by Dr Whewell as an historical term, now superseded by the more general and comprehensive phrase, Moral Philosophy. On the other hand Professor Maurice, while he looks on the solution of cases of conscience as impossible, applies the name of Casuistry to that subjective side of Moral Science, which enforces the claims of the personal "I," the individual conscience. This subject, though highly important in itself, does not seem to me, whether in history or by etymology, to answer to the name. But I think it may be transferred, by a moderate license, from doubtful and disputed questions in the details of Moral Duty, to Controversial Ethics, or the attempt to gain clear and firm convictions on those great questions, which give birth to rival schools of ethical teaching, and have perplexed and divided the judgments of moralists for thousands of years.

The present work treats of three topics, all preliminary to the direct exposition of the first of these three main subjects, Moral Philosophy. These are the Certainty and Dignity of Moral Science, its Spiritual Geography, or relation to other main subjects of human thought, and its Formative Principles, or some elementary truths on which its whole development must depend. In the coming year I propose to myself to enter on the second subject, or Controversial Ethics; by a review, first, of Modern Utilitarianism, as expounded by Paley and Bentham, and recast by Mr Mill into a different form; and next, of modern Cambridge Ethics, represented by the Discourse of Professor Sedgwick, and the writings of my three eminent predecessors. In Mr Mill's review of Professor Sedgwick these two schools came first into direct collision, almost forty years ago; and now in the present year, within a few months, these distinguished writers have both passed away. The harsher sounds of controversy should be stilled, and only its grave and gentle utterances be heard, over the recent graves of the dead.

I have ventured to append to these Lectures a college essay or declamation, delivered in Trinity College Chapel in December 1833, or just forty years ago. I believe that the thoughts it contains, however youthful the style, are seasonable and important at the present hour. They secured at the time a favourable notice from Dr Chalmers and some other distinguished men. But I reproduce them here for a double reason. They are a pledge that the views held in the present volume, and others which may follow, are no hasty product of recent study, but convictions early formed, and deepened by all the study and reflection of so many years. Its date is just one year after Professor Sedgwick's Discourse, and a few months after its publication. It may thus be taken as one further sign of a reaction against the selfish and utilitarian school of ethical teaching, which had then set in, alike among older and younger members of the university, and which has continued to the present day.

I commit this work, as a small sheaf of first-fruits, to the candid and forbearing perusal of learned readers; but still more to the blessing of Him, without whom nothing is strong, nothing is holy, the only Fountain of moral insight and true wisdom, the uncreated and eternal Goodness, in whom all truth dwells in its perfect fulness, from whom its streams proceed, and to whom they return, after watering the wide universe of moral being through which they flow.

TRINITY PARSONAGE,
CAMBRIDGE,
September, 1873.

LECTURE I.

THE TRUE PLACE OF MORAL SCIENCE.

IN entering on the duties of the Knightbridge Professorship, I have endeavoured, in my Inaugural Lecture, to point out the great importance of the systematic study of Moral Science in the days in which we live. It is natural to begin this first course of Lectures with some remarks on the main outlines of the subject, and the order of treatment I hope to pursue. This seems doubly needful, when I succeed to writers so eminent as Dr Whewell and Professor Maurice. They have left behind them, besides four volumes of published Lectures, a full and detailed History of Moral Philosophy, ancient and modern, and a systematic treatise on the Elements of Morality. And still they vary considerably from each other in their construction of the titles by which this Professorship is defined, and in their methods of ethical exposition.

The original subject, by the deed of the founder, Dr Knightbridge, is "Moral Theology, or Casuistical Divinity." But when Dr Whewell, on his election in June 1838, revived the office from a long and deep slumber, he introduced in his first Lecture the name

"Moral Philosophy", as well adapted, in his view, to express "the substitution of a newer form of science, full of life, hope, interest, and solid truth, for older and more imperfect speculations." "I shall reckon," he says, "on the implied sanction of the University, in considering myself as Professor of Moral Philosophy, a branch of study of which a Professorship exists, I believe, in every University but our own." The sanction implied at first, was afterwards expressly and publicly given, when the University made a later addition to the original endowment.

Dr Whewell began with a course of Lectures on the English Moralists from Hobbes to Paley, Gisborne, and Price, at the close of last century. To these, when published, six others were added on Bentham's works; and in a second edition, in 1862, fourteen others, which began with Plato and ended with Coleridge. His second and main work, published in 1846, was the "Elements of Morality," a full and systematic treatise on Moral Science. It was followed, the same year, by eight "Lectures on Systematic Morality," and appeared anew in 1864, shortly before his death, in an enlarged and revised edition.

The Third Book of the Elements includes a full treatment of Ethics in their religious aspect, or Moral Theology. But Casuistry, in Dr Whewell's view, holds quite a secondary place in Christian Morals. It refers, he says, to "questions of human conduct, in which conflicting duties, or obscurity in the application of moral rules, perplex and distort the faculty which judges of right and wrong." It is thus "neither the main part" of the general subject, "nor that from which it can with propriety derive its name."

Prof. Maurice, on the other hand, in his opening Lecture, seems to regard Casuistry, Moral Philosophy, and

Moral Theology, as three distinct and successive stages of Moral Science, which fitly express its personal, social, and religious elements. He appears further to assume that these titles of the office are a guiding light, which the University itself has supplied, to fix the limitations of the main subject, and the due and proper order of its exposition. He had published, twenty years before his election, an instructive treatise on the history of Mental and Moral Philosophy, and it has since appeared, in a revised and enlarged form, as his latest work. It was natural for him, then, to abstain in his Lectures from this branch of the main subject, and to confine himself to a more direct treatment of moral questions. His first course, accordingly, was on the Conscience, under the title of "Lectures on Casuistry," and his second on Social Morality. Had his life been spared, we may assume that a third would have followed on Moral Theology. But he remarks that he had found it impracticable to maintain entire distinctness, and that Moral Theology had really intruded into both. "It must be so," he observes, "for any one who discovers, beneath the conscience which testifies of our personal existence, and beneath all the order of human society, a Divine foundation."

Dr Whewell, then, has unfolded Ethical Science mainly on its objective side, as a connected and very extensive scheme of thought, under the heads of Springs of Action, Morality, Religion, Rights and Obligations, Polity and International Law. But in such an outline, however ably filled, there may be some danger of the conscience finding itself oppressed, like David in Saul's armour, by a panoply of solid and laborious thought, too heavy for its weakness to sustain. A treatment less complete and exhaustive, but appealing more directly to the spontaneous emotions of the heart, like arrows from the bow, or stones from the sling, must therefore have been desirable at least as a supplement for common minds.

Prof. Maurice, accordingly, took up in his first course the subjective side of Ethics. He sought earnestly to bring out into full relief the "I ought" of conscience, to protest against theories which would resolve it into the mere dread of human punishment, and to summon it, by a direct appeal to the consciousness of its own supremacy, to the pursuit of high and noble aims. An awakened conscience, fully alive to the claims of duty, which looks up with reverence to a law it cannot alter and is bound to obey, is the first essential of true morality, the only genuine passport to the temple of ethical science. Where this is absent, learned speculations on moral theories, and on schools of ancient and modern thought, become immoral trifling, bewildering to the reason, and deadening to the heart.

But while the subject of these Lectures is, I conceive, a most suitable and needful supplement to Dr Whewell's more systematic work, I do not think that they are "Lectures on Casuistry," either in the sense of the founder, or in the well-known and historical meaning of the name. The earnest inculcation of thorough conscientiousness in all cases, and under all sacrifices, is indeed of vital and almost supreme importance to the moral student. It is the starting-point of genuine progress, the living sap in the tree of Ethical Science. But Casuistry in its proper sense, or the provision of rules for the guidance of such a conscience in doubtful and obscure cases, is widely different. One is like the root of the fruitful olive-tree, the other is the gleaning berries on its outmost branches, when the harvest has already been gathered in.

The threefold title of the Professorship may be adopted, I think, with less violence to the real distinctions of the subject, in another way, without disturbing their order, or using Casuistry in a sense wholly diverse from its historical meaning. Moral Theology, instead of being included, as in Dr Whewell's work, midway in the course, should be reserved, as the highest division, for the crown and climax of the whole. Philosophy deals with the conclusions of human reason, aided and enlightened by Revelation, but acting still within the limits of that Divine appeal—"Yea, and why even of yourselves judge ye not that which is right?" Theology deals with the character and works of God, as made known to us more clearly and fully by Divine revelation. And thus Moral Theology is the meeting-place and border province of these two kingdoms of thought. It deals with the religious aspect of Moral Philosophy, and the ethical aspect of Revealed Religion. It binds these together in a higher synthesis, where reason is ennobled and purified by faith in a Divine message; while faith itself, cleansed from mere superstition, offers to God a reasonable service, and sees light in the light of heaven.

Moral Philosophy, thus defined, must include the subjective and objective side of Ethics, the enforcement of the claims and supremacy of conscience, and the whole range of social morality; or, in one word, all the direct and positive teaching of Moral Science. What place, then, is left for Casuistry? In the strict and proper sense, Dr Whewell justly regards it as a very subordinate branch of the general subject. The difficulties, also, to which it relates are more likely to seek and find their solution from the pastor or the friend than from academical lectures. And Professor Maurice, even while he retains the term, seems wholly to abolish the science or art itself, since he holds that attempts to lay down rules for cases of conscience "only leave those cases more unsettled than ever." But if doubts and questionings, in the present day, are mainly transferred from curious and involved cases, where the path of duty is obscure, to the very foundations of morality, we may come near to the true design of the founder by retaining the title in a somewhat varied sense. Casuistry will thus refer to the polemical aspect of moral questions, and deal with those controversies and disputes which tend to cloud and perplex the minds of students, and are constant attendants on every imperfect stage of a progressive science.

I propose, then, to arrange the general subject under these three main divisions: 1st, Direct Ethics, or Moral Science in its relation to other sciences, its fundamental principles, and the wide range of personal and social obligations, including some first lessons of religious faith and duty; 2ndly, Controversial Ethics, answering nearly to Casuistry, or the inquiry into the debateable ground of Morals, and the discussion of rival and conflicting systems; and, 3rdly, Moral Theology, a wide and interesting, as well as difficult, field of thought. For here an awakened conscience and intelligent Christian faith act and react on each other. The conscience tests pretended revelations by laws of its own, while it submits reverently to the authority of a Divine message. It can thus learn new facts and lessons of God's moral government, grow more fully instructed as to its own nature and the duties and hopes of man, as responsible to his Maker, and walk in the light of a teaching higher and even purer than its own. Here, also, an intelligent, but sober and reverent, faith brings its unsolved and perplexing problems, in messages that claim a Divine origin, to be examined in the light of a purified reason, and to be solved even here, when the answer is possible, for our present comfort, and the due strengthening of its own powers. But when such a solution is too hard for present attainment, it is the further office of Moral Theology to remind us of the limitations of our moral insight, while we see "as in a mirror, darkly"; and to encourage us to wait hopefully, and not impatiently, for clearer light; whether the blessing may be given in the world's eventide, or only in that brighter morning, when the day shall break on the immortal spirit, and the shadows of the night shall flee away.

These few remarks on the conflicting definitions of Casuistry by the two most eminent of my predecessors, and their differing arrangements of the whole subject, seemed needful to explain and justify the order I propose to follow. My purpose, in the present course, is to treat of some first principles of Direct Ethics, or Moral Philosophy in its narrower sense, under these two heads: its relation, as in a moral geography, to other branches of human knowledge, and its fundamental principles, or the courses of solid masonry on which it rests below. And my first subject will be the True Place and real dignity of Moral Science.

There is a school of thought, in our own days, which, under the boasted name of Scientific Progress, would wholly abolish all Theology, and reduce Morals to a dependent and precarious existence, in which it becomes the vassal and slave of Physics or Human Anatomy. And its strength, in popular esteem, has been derived, in no small measure, from its attempted classification of every department of human science. The instinctive craving after unity, never wholly asleep, has been wakened into new and more intense activity by the discoveries of the last hundred years. Many run to and fro. The bonds of commercial intercourse are interlaced and strengthened. Nations learn thus to look beyond themselves, and recognize dimly a wider and larger brotherhood of all mankind. The same influence extends to the mutual relations of the sciences themselves. The various branches of intellectual speculation and thought borrow from each other, and melt into each other, more and more. Thus Astronomy borrows from Mechanics, Optics, Electricity and Magnetism, and even of late from Chemistry, the means for its further and wider progress. Geographical research becomes more exhaustive, and geological inquiry deeper, more profound, and more various than before. Both of these alike borrow largely from Zoology and Botany, and lend to these sciences rich and copious materials in return. Heat is identified with motion, and becomes a subtler branch of Mechanics. The magnetism of the earth has its changes linked with solar phenomena. Deep ocean soundings, made by new refinements of instrumental skill, modify geological theories. The same craving for comprehensiveness and unity reveals itself in antiquarian pursuits, and modern researches into the history of language and of race. It seems as if the limbs of Science, torn asunder by ignorance, and violently scattered, were striving to come together, and clothe themselves with sinews and flesh once more, and thus to form one living, united, and harmonious system.

In such a period of thought, any school which professes to exhibit all the sciences in their orderly succession, and to ground on that arrangement special theories and deductions of its own, wields an engine of aggression on all rival systems of no common power. The combat is

THE TRUE PLACE OF MORAL SCIENCE. 9

like that between a regular army with its well-disciplined battalions, and guerillas or volunteers without discipline or concert, spread out in loose array.

Such a theory has arisen, it is well known, in French Positivism, and, variously modified, has many disciples or admirers among the cultivated classes of our own land. But such a doctrinal system, which destroys and proscribes all religious faith, and degrades morality into a cerebral secretion, a blind necessity, or a pleasure-seeking prudence, can only be met and overthrown by one better and nobler; by a system which retains whatever is true and sound in the lower fields of thought, but includes in its larger geography those sacred heights and mountain-tops of human science, which deal with moral and spiritual objects, and pierce into the skies above, while they command the widest and most comprehensive views of the peaceful valleys below.

Let us begin, then, by endeavouring to fix clearly the true place of Ethics in the wide range of human science. And here the scheme of Lord Bacon, in his *De Augmentis*, supplies a basis, capable indeed of no slight improvement, but from which the remarks of Locke, and the fuller system of Comte, are really a retrogression and decline, rather than an advance to clearer insight, and a more perfect and comprehensive arrangement.

The first division of human knowledge, Bacon has well observed, must be drawn from the threefold powers of the rational soul, which is the seat of knowledge. History belongs to the memory, Poetry to the fancy or imagination, and Philosophy to the reason. But here two further remarks naturally arise. First, this division is common alike to every branch of human thought, just as breadth, in Geometry, coexists with the other dimensions of space. It seems thus more natural and convenient, instead of separating History and Poetry altogether from the kindred branches of Philosophy, as Bacon has done, to make this distinction subordinate to the main objective divisions of Human Science. And next, in the case of Natural Philosophy, the registration of natural phenomena, which answers to history, and the invention of hypotheses and development of their consequences, which correspond to poetry, are properly quite subordinate to the branch of science to which those phenomena and hypotheses belong. They form its materials, and supply the means for its further progress. Tables of observations, records of observed facts, registered stellar phenomena, and the successive theories of Ptolemy and Copernicus, of Kepler and Herschel, are thus the inseparable adjuncts of astronomical science.

Science or Philosophy, as Bacon has well observed, receives its primary and fundamental division from its threefold object, "Deus, natura, homo." "It is convenient," he adds, "that Philosophy be divided into three sciences, concerning God, nature, and man." But when he has laid this clear and simple foundation, he proceeds at once, rather strangely, to deviate from the only true and natural arrangement. He distinguishes inspired Theology wholly from Philosophy, and reserves it for the last place at the close of his treatise, as "the haven and sabbath of all human contemplation." And yet he places Natural Theology, as a separate branch, first in the three main divisions of Human Philosophy.

Such a scheme has plainly a double fault. It severs Natural and Revealed Theology as widely as possible, making them form the two extremes of the whole system. But they are plainly mere subdivisions of the same grand subject. The contrast implied in the names does not extend to the truths themselves, but refers wholly to the different means by which our knowledge of them is supposed to be attained. And next, his plan involves a disturbed order of the three divisions themselves, which are not allowed to form either an ascending or a descending scale.

The truer plan of arrangement, in this respect, which M. Comte has justly followed, is that of an ascending series. Natural Theology, in Lord Bacon's outline of sciences, should plainly be reserved for the third place, near the close; so as to prepare the way, by a gradual ascent, for the fuller and higher teaching of Divine Revelation. Nor should this be excluded from Philosophy. It ought rather to be included in it as its highest and noblest portion. For the genuine pursuit of wisdom, which is the meaning of philosophy, can never rest short of the vision of the Only Wise, or that "knowledge of the Holy" (Prov. xxx. 3), which alone is true understanding and "life eternal."

. Again, that "prima philosophia," which Lord Bacon places as the common stem of the three main divisions, seems to have no real claim to this high position. So far as it is real, and not a mere play on words, or on the analogies and ambiguities of human language, it belongs to some aspect or other of Natural Theology, and the relations between the First Cause, the Creator, and the whole universe of created things.

The first and main division, then, of Philosophy or Science is threefold. Its first and lowest portion is Natural Philosophy; its second, Humanity; its last and highest, Theology, both Natural and Revealed. This, too, is the very order implied in the system of our ancient universities. It is far more full, harmonious, and complete, than the maimed and imperfect substitutes which have been proposed or invented in modern times.

Natural Philosophy is divided by Lord Bacon into Speculative and Operative, and the Speculative into Physical and Metaphysical. The former of these is said to deal with the efficient cause and matter,] the other with the final cause and form. Again, the Physical is divided into three subjects, the Principles of things, the World, and the Variety of things. It is further divided into a doctrine of Concretes and Abstracts, the first having the same divisions as Natural History. The Metaphysical is divided into the doctrine of Forms and of Final Causes.

Operative Physics, again, are parted into Mechanics, and Magic in a revised sense of the term. This, in Bacon's view of it, seems nearly to agree with the modern science of Imponderables, since he defines it to be "a practical knowledge of the more secret powers and subtle influences of the natural world." Last of all, Mathematics, both pure and mixed, are made a separate appendix of the whole.

Such an arrangement of Natural Science, it can hardly be denied, leaves on the mind a vague and confused impression. From the objective principle, clearly laid down at first, we are thrown back on subjective divisions of a wholly different kind, which are made to depend, in part, on the abstractions of Greek philosophy. What is called Operative Natural Philosophy really consists in the application of natural knowledge to the wants of mankind, and thus properly forms a part of the higher subject of Humanity. Lord Bacon seems to have known little of Mathematics, and, like Sir W. Hamilton, held them in light esteem. Instead of being made an accidental appendix of Physics, they ought rather to come first, as the intellectual starting-point and basis of the whole scheme. They are lowest, indeed, in dignity, though first in order of ascent, but still claim a very high importance, because their truths underlie every branch of Material Physics, and may thus be justly held to be the foundation of the whole.

Mathematics, or the Science of Number, Space, Motion and Force, will thus hold the first place in the natural arrangement of the Physical Sciences. Next will come Uranology, or the knowledge of the heavens, and all the worlds they contain, whether suns, stars, planets, satellites, or meteors, star-dust, and nebulous matter, spread through the depths of space. Third in order will be Ecumenology, or the knowledge of the habitable world, a subject in outward extent far more limited than the last, but also far more accessible, and thus capable, practically, of a fuller development, and of more various subdivision. It includes three main classes of objects—lifeless matter, plants, and animals or living things. The first may be called Hylology, and is either analytic or synthetic. Physics, or Analytical Hylology, will include Solid Mechanics, Hydrodynamics, Pneumatics, the doctrine of the Imponderables—Heat, Light, Electricity, and Magnetism—and analytical Chemistry, or the determination of the laws which distinguish the main classes of lifeless things. Geognosy, or Synthetic Hylology, will include Physical Geography, Hydrography, Meteorology, and Geology; with two appendices of Palæontology, or the determination of past changes, and the probable state of the Earth in every former age; and a corresponding science, still nameless, and beyond the range of present human foresight, which would include the prediction of all future changes in distant ages still to come.

Passing over the two higher divisions of Natural Philosophy, Botany and Zoology, we come to a higher subject, which holds the middle place in the ascending series, and to which the words of Pope will apply—
The proper study of mankind is Man.

The leading division in the *De Augmentis* of Lord Bacon has here been reproduced by M. Comte under different names. The Philosophy of Humanity and Civil Science, in the former, answer very closely to Biology and Sociology. The first is parted by Bacon into a science of the Body and of the Soul. A third division is added, on the State of Man, or his Personality, and the union of body and soul. This, again, is made to include two parts, a doctrine of the Miseries, and one of the Prerogatives of Man. The Science of the Body is parted into four divisions, relating to health, beauty, exercise and pleasure, or a science of Medicine, Cosmetics, Athletics, and Hedonics. The Science of the Soul is parted into one of the Substance and Faculties, and another on the use and objects of those faculties. This latter is divided into Logic and Ethics. Ethical Science is parted into a doctrine on the Standard of Good, and another on the Culture and Guidance of the Mind; and this last into three portions, on Character, on the Affections, and on Moral Remedies.

Civil Science, again, is ranked under three main divisions. The first is the Science of Conversation, or Social Intercourse; the second, of Business or Occupation; and the third, on the Republic or Empire. To this brief outline are added two supplements, on the fountain of social right, and on enlargement of the bounds of empire.

A simpler arrangement, I conceive, may be based on the ascending scale of human faculties or powers, and will thus be in closer harmony with the main objective division of all science into Natural Philosophy, Humanity, and Theology.

Humanity, or the Science of Man, in its widest sense will include the knowledge of Human Action, Speech, and Thought. Human Action will admit a fourfold distinction, with reference to the individual, the outer world, the family, and the state. These might perhaps be conveniently styled Autonomics, or the discipline and culture of man's own person and bodily frame; Geonomics, including agriculture, horticulture, zoonomy, or the culture of animals, and navigation; Economics, or the science of domestic and family life; and External Politics, or the science of human action, when men are gathered in civil societies of various kinds. The Science of Speech will include Grammar, Lexicography, Oral and Written Language, Semeiology, or various modifications, used as signs of thought, such as Hieroglyphics, Stenography, Secret Writing and Telegraphy; and the higher branches of Logic, Rhetoric, Dialectics, Education, Jurisprudence, Literature, and Public Worship. The Science of Thought will include Mental Philosophy in all its various aspects, and is either Analytic or Synthetic. The first refers to the different faculties and powers in each individual, and includes a doctrine of Perception and Sensation, or the relations of the mind to outward nature; of Reflection or Self-knowledge, of Sympathy and human fellowship, and of Religious Faith, aspiring to things unseen and eternal. The Synthetic Science of Human Thought is that which deals with the various characters and classes of mankind, and all the diversities of sex, age, race, intelligence, culture, and the countless varieties of human life and feeling, of social and religious thought.

But a Science of Humanity, when pursued within these limits, does not satisfy the conditions of scientific completeness. For it deals wholly with the actual, and not the ideal. It contemplates Man as he has been or now is, and not as he might be or ought to be. But Man is conscious of powers of choice, on the use or abuse of which his happiness very mainly depends. He is not a mere tool or engine, set in motion by external powers, over which he has no control. He has a knowledge of good and evil, of evil which he seeks to avoid, and of good which he dimly seeks after, and longs to attain. He feels himself capable of progress and improvement, or of degeneracy and decline. Herein he feels himself to differ widely from lifeless matter, with its laws which it must obey, and even from all the lower animals, though these are gifted with wonderful instincts and powers of spontaneous motion. And thus there remains a higher field, beyond and above all those branches of Human Science which have now been briefly indicated, and coextensive with their whole range. Man's nature is twofold. It includes the consciousness of actual powers and capacities, and the dim perception of a high and noble ideal, attainable, but not yet attained. This is well expressed in the often-quoted lines—

> Except above himself he can
> Erect himself, how mean a thing is Man!

Humanity, then, since it refers to a being far higher in its powers than lifeless matter, or mere animal instinct, and still far below Divine perfection, resolves itself necessarily into two main divisions. The first is Actual Humanity, or the knowledge of Man such as experience proves him to be, in his various relations to nature, to his fellow-men, and to the Unseen Power on whom his being depends.

The second is Ethics or Ideal Humanity, the knowledge of that high standard of perfect action, speech and thought, below which men may fall continually, and too often with a deplorable and melancholy contrast, but towards which they are bound ever to aspire; and, aspiring towards it with earnest desire and effort, may hope for its fuller and fuller attainment.

Ethics, then, in one word is the Science of Ideal Humanity. It sets before us Man, not as he is, but as he ought to be. It implies a standard of right and wrong, which does not depend on the actual state and conduct of mankind, and is not fixed by past experience, but which shines out amidst the storm-clouds of human passions and vices like a rainbow of hope and promise, pointing onward to something bright, excellent and glorious, not yet attained. This science of Ideal Humanity is the true mainspring of all human progress, which really deserves the name. And it forms also the natural transition to the best and highest field of human thought, Divine Theology. The connection is no mere result of fancy, or philosophical reasoning. It is inwoven into the very texture of Christian faith. For this is the grand "mystery of godliness," on which the whole fabric of the Christian revelation depends, that the ideal Man is no other than the Incarnate Son of God.

A clear view of the main outlines of human knowledge, and of their mutual relation, will thus enable us not only to ascertain the true place, but to maintain the dignity, of Moral Science. The arrangement of Lord Bacon is in this respect very faulty and imperfect. It would lead us to suppose that Ethics are a mere subdivision of one subdivided branch of the doctrine of Humanity; that they come nearly midway in its course, and form hardly one-tenth of the whole. But this is wholly different from the real truth. They constitute a vast and wide field of thought, conterminous with the whole range of actual human knowledge. Throughout all the wide expanse of human interests they prescribe a standard of perfection to the actions, words, and thoughts of men, towards which they are bound unceasingly to aspire.

Ethical Science is no mere product and corollary of man's past experience. It is rather its needful antidote. Its motto is evermore "Excelsior." It never permits this standard to be torn from the staff, and trailed in the mire of human corruption. It calls unceasingly on the corrupt and the impure to awake and arise. Amidst the strife of parties, the speculations of false philosophy, and the seductions of sensual vice, its clear and solemn voices are heard continually in such utterances as these: "Thou shalt not follow a multitude to do evil." "Speak to the children of Israel that they go forward." "Whatsoever things are lovely, whatsoever things are of good report, if there be any virtue, and if there be any praise, think on these things."

This view of the true place and correct definition of Ethics as the Science of Ideal Humanity agrees closely with some striking remarks in Professor Grote's *Examination*, a posthumous work, full of careful and suggestive thought.

"There is no moral logic which will teach us to conclude what should be, in the great features of it, from what has been, and what is. If we do so conclude, it is in a manner which destroys all our moral being. Man has improved as he has, because certain portions of his race have had in them the ideal element, have been unsatisfied with what to them at the time has been the positive, the matter of fact, the immediately utilitarian; have risen above the cares of self and of the day, have been imaginative in thought, enterprising in action, deep and earnest in feeling....If what Man's experience teaches him is to give up the imaginative, the deep and unsatisfied thoughtfulness, the desire to penetrate to the reason of things, the hopefulness of becoming a worthier and higher creature; if it teaches him to be content with the idea of knowledge as the registering of facts, and as what, rightly used, may benefit his material condition—he will, I think, cease to improve. If he had acted on this principle from the first, he would never even have begun to improve."

Two distinct charges may be brought against Ethical Science, as thus defined. The first, that it must be unpractical, dreamy, and Utopian. The second, that it is barren and inoperative, confined to a few popular truths, and incapable of real progress.

Professor Maurice has remarked, in the opening of his first lecture,—" If the moral teacher adopts the distinction which is sanctioned by one of the ablest and most accomplished of his class—that his business is with what ought to be, that of other students with what is, can there be a clearer or fuller confession that he means to leave the actual world for some other world which he has imagined?" And again, in the third lecture, after naming Sir J. Mackintosh as the source of the quotation, he resumes, " The distinction was plausible in itself, even without considering the authority from which it proceeded. Yet if we accepted it, Ethics seemed transferred from the real world in which we dwell to some other imaginary world. In this case I am sure we should get no serious attention for them in this busy, practical age. Dismissing, therefore, this opinion, without examining what might be the arguments for it, we asked whether there was no other difference between this study and those with which we are engaged elsewhere....The Moralist cannot be less immediately occupied with that which is, with existing facts, than any physical student. His business cannot be carried on in some distant Atlantis, nor can he be engrossed in the search for one."

In passages of this kind it is not always easy to know whether my predecessor merely describes the probable feelings and impressions of others, or adopts them for his own. But it seems a cause for regret that there should be no formal censure of an inference, which, if actually drawn, would be one of the most inexcusable follies into which this busy, practical age could possibly fall. Ethical Science may be safely neglected, and in such an age is sure to be neglected, if it deals, not with what is, but with what ought to be! Now the true sense of the definition is plain. Neither Sir J. Mackintosh nor any other moralist of common sense could ever be supposed to mean that the business of Moral Science is to quarrel with God's constitution of the universe, or to copy the traditional blasphemy of Alphonso of Castile, who said that he could have taught the Creator how to frame a much better and more perfect world. Moral Science, it simply affirms, teaches us how men ought to act, not how they have acted in time past, are acting now, or may, with more or less likelihood, be expected to act in time to come. It deals, of course, with the actual conditions and circumstances of human life. But still its proper work is to teach what human actions ought to be, not what they have been or are, in all the countless relations they fulfil to each other, and to the great Author of their being.

Such is plainly the meaning of Sir J. Mackintosh, in the

words of his brief definition. They involve no confession, whether clear or obscure, of a purpose to forsake realities, and to speculate principally on some distant and imaginary world. Should a busy, practical age turn away from them, and justify its conduct by such a plea, our first duty, as moralists, is to expose the misconception, and point out the mischievous folly of the practical result to which it leads. It is just in proportion as any age is really practical, that the maxim has the highest claim on their notice and their reverence. For it reminds these busy, practical men, that the world in which their lot is cast is not as it ought to be, because they themselves, the moral agents by whom it is peopled, are not doing as they ought to do.

There is a certain sense, it is true, in which the maxim does labour to transport men from the actual to an imaginary world. Only that other world is no fabled Atlantis, no unknown planet, governed by laws which human fancy has devised, and forming a part of some unknown system. It is our own world, changed and transfigured by a moral renovation, when a pure and noble ideal of thought, speech, and action, shall once have been deeply and abidingly implanted in the hearts of men. Its nature has been well and clearly expounded by our Christian poet,—

> O for a world, in principle as chaste
> As this is gross and selfish! one in which
> Custom and prejudice shall bear no sway;
> That govern all things here, shouldering aside
> The pure and modest Truth, and forcing her
> To seek a refuge from the tongues of strife
> In nooks obscure, far from the ways of men!

Morality is thus an intensely practical science, and

claims the most eager and earnest attention in a practical age of busy men, for this very reason, that it exhibits before their eyes a lofty ideal of right conduct in all the relations of life, which it is our bounden duty, every day and every hour, to strive more nearly to attain. It admits, in accomplished facts, or in the voice of numbers and the clamours of a multitude, no excuse whatever for selfishness, vice, and crime, but reminds men solemnly what they ought to be, and the high standard they ought to keep ever full in view.

But ethical study, if it escapes the charge of being dreamy, unreal, and Utopian, is exposed to a kindred reproach, that it is barren in all results, stationary and unprogressive. Thus we read as follows in a work of some reputation:

"Though moral excellence is to most persons more attractive than intellectual, it is far less active, less productive of real good. The efforts of the most active philanthropy, the most disinterested kindness, are short-lived, and rarely survive the generation which witnessed their commencement....There is nothing to be found in the world which has undergone so little change as the great dogmas of which moral systems are composed. To do good to others, to love your neighbours as yourself, to forgive your enemies, to restrain your passions, to honour your parents, to respect those who are set over you— these, and a few others, are the sole essentials of morals. But they have been known for thousands of years, and not one jot or tittle has been added to them by all the sermons, homilies, and text-books, which moralists and theologians have been able to produce."

Moral Science, if these strictures are just, consists of little more than half a dozen sentences, known for thousands of years, and containing maxims which very few have cared to practise. These maxims are incapable of any real addition, and have also exercised a very slight influence on the welfare and happiness of mankind. Physical Science, it is affirmed on the other hand, consists of an immense body of truth. It is cumulative and progressive in its nature, and has grown constantly from age to age. And thus we may reasonably hope that it will conduct mankind by a surer path than mere moral teaching, and with a more effectual guidance, to some distant goal of social happiness and well-being.

Now if it be meant simply to affirm that the first principles of Moral Truth have been known long ago, the same is clearly true of arithmetic, mechanics, and geometry. Yet their progress has been as real, and almost as conspicuous, as that of the Physical Sciences which depend upon them. On the other hand, if it be affirmed that simplicity and permanence, in general laws, forbid any development of the various results to which they lead, the facts of modern astronomy offer the paradox a decisive refutation. The moral law, which enjoins the love of our neighbour, is hardly more brief, and certainly not much more simple, than the law of gravitation. Yet the ablest analysts and geometers, for two hundred years, have tasked their powers to the utmost in tracing out the results of Newton's great discovery. And still they are very far from having exhausted the mine of its intellectual treasures. In binary stars, in asteroids and planets once unknown, in comets and meteorolites, in lunar disturbances, and tidal retardation, new discoveries reward their efforts from age to age. And after all their labour, the problem of three bodies, one of the simplest that can be proposed, defies their attempts at a perfect answer, and can be solved only under favourable conditions, and by gradual and successive approximation.

Why, then, should Moral Axioms be less fertile than the Laws of Physics in the results to which they lead? The field is higher and nobler, and the capability of large development, on every ground of reason, is just the same. If mankind at large were half as zealous in the pursuit of moral excellence, as astronomers have been, since the days of Newton, in their calculation of attractive forces, and their practical study of the heavens, a thousand years would not suffice to exhaust the various development of the great laws of social duty, or bring to a close their progress in moral insight, and their successful labours of thought, and practical endeavours in this higher field.

Once let us see clearly the true place of Morals in the grand series of the sciences, and these censures aimed against them by authors who may be clever in their own pursuits, but are sorely wanting in moral discernment and true wisdom, will drop away like withered leaves, and disappear. A science, of which the very aim and purpose is to discover and enforce the true ideal of right feeling and right action for every moral agent, and for each individual of mankind, cannot possibly be barren and worthless, or devoid of practical power. It must be, from its very nature, of high and inestimable worth. Spurious counterfeits, indeed, may be not only unprofitable, but most mischievous. But surely a lofty and pure ideal of thought and action can never be set before the eyes of men, even in the most corrupt age, and the most degraded state of society, wholly in vain. If such truths, plainly taught, and such high aims, held up before their eyes, were to be barren of all results, then assuredly, in the words of Milton,—

> The pillared firmament is rottenness,
> And earth's base built on stubble!

It is quite conceivable, however, that the mode of operation of such truths, and the amount of influence they exercise, may wholly elude the notice of keen-eyed worldly men, or even of clever essayists and philosophers, intent on physical research, or buried in the strife of parties in the political world. There are tens of thousands who feel the steep ascent of a hill-side, or even the weight of some slight burden they carry in their hand. But who has felt, or feels at this hour, that mighty force of solar attraction, which has been unceasingly at work, on the largest scale of dynamical energy, through successive ages? Yet this alone has kept our earth stedfast in its orbit, and hindered it from losing itself long ago, with all its inhabitants, in outer darkness.

When the stern and grand old prophet stood on Horeb, on the mount of the law, it was not in the earthquake and the fire, the wind and the storm, that he recognized the special signs of the Divine presence. It might have been known and felt there also, but it was found and felt chiefly in the still small voice alone. In nature those influences are the most penetrating and powerful which escape the gaze of the superficial observer. There can be no doubt, either to calm students of human history, or to firm believers in Christianity, that the practical influence of great moral truths, in all past ages, has been grievously hindered, and often neutralized, by the moral dulness, or the vicious and sinful perverseness, of the greater part of mankind. They have been too often like dews alighting on the rock, or seeds that are sown on a barren and ungrateful soil. But it would be a fatal and immense error, on this account, to imagine that their publication, their ceaseless iteration, and earnest enforcement, whether by moralists or divines, have been wholly in vain. Whenever the thoughts are turned to them, the dullest conscience is in some degree stirred and aroused. The coldest heart is either touched with a secret pang of remorse, or kindled into dim and secret longing after a higher life. The weak and irresolute are awakened from the sleep of useless indolence, and nerved for conflict with temptation. The trance and stupor of sensuality is disturbed or broken. A voice—What meanest thou, O sleeper? awake and arise! startles the truant conscience in its guilty wanderings. A breath of heavenly life seems to visit and breathe on the moral being, like the change which comes on all the face of nature in the first days of spring. Such emotions, indeed, when left to themselves, may quickly expire, and the coldness become greater, the darkness deeper than before. Some higher power is needed to sustain the awakened spirit, to keep it from relapsing into double apathy, and to guide its steps along the steep hill-side of heavenly truth. When the brightness and beauty of a high moral standard has dawned on the feeble and tempted spirit, the first impulses of awakened thought need to be sustained by prayer for Divine help, and the hand to be stretched out eagerly, to meet the proffered succour of heavenly grace. The parting words of Milton in Comus will then be found to be no mere utterance of a sportive fancy, but the veiled expression of the deepest philosophy, and of the highest lesson of Christian faith —a faith and a philosophy far more profound than modern theories for manufacturing some miserable semblance of a conscience out of the transmuted instincts of the ape or baboon:—

> Mortals, who would follow me,
> Love Virtue, she alone is free:
> She will teach you how to climb
> Higher than the sphery clime:
> Or, if Virtue feeble were,
> Heaven itself would stoop to her!

LECTURE II.

THE CERTAINTY OF MORAL TRUTH.

ETHICS is the Science of Ideal Humanity. It is no mere subdivision of one secondary and limited province in the wide domain of human thought. It is rather like that blue firmament, which is above us and around us wherever we go, and encompasses the lofty mountain summits and the lowly valleys of earth on every side. Its aim is to teach men, in every various field of thought and action, what they ought to be, how they ought to live, and what they ought to do. It claims to preside over all their converse with the world of nature, and the conduct of their inner life, their relations to their fellow-creatures, and their duties of prayer, praise, trust and worship, towards the Supreme Creator.

In width of range it must thus be coextensive with all the various and countless fields of human thought, speech, and action. In dignity it rises above them, and surmounts them all. They set before us man as he is; Ethics, man as he ought to be. They unfold his actual powers, habits, tendencies and dispositions. The science of Ethics sets before him a pure and lofty ideal, of good which he ought to seek, of moral beauty and perfection which he should ever strive to attain. Its lessons, when fully learned, show him how far he has come short of this high standard, which his own conscience, when thoroughly awakened, is compelled to approve. And thus it leads him by the hand to that footstool of revealed mercy, where heaven stoops to the feebleness of human virtue, or even to the degradation of human vice and folly, and with sovereign power raises up the lost and guilty to holy blessedness without measure and without end.

But here, at the entrance of our inquiry, a great stumblingblock lies in our way, and needs to be removed. Have Ethics any just claim to be a genuine science? Is certainty, on moral questions, possible or attainable? Do we not rather enter here on a dark path, a thorny jungle of barren strifes of words, and tedious disputations "never ending, still beginning," and renewed with a wearisome and fruitless pertinacity from age to age?

The past history of mankind at large, and the known course and cycles of moral speculation, tend to strengthen these doubts, either of the reality, or else of the attainableness and practical worth, of ethical science. The fact is undeniable, which Locke has made so prominent in his reasoning against innate ideas (Bk. I. ch. 3, § 9), that a low and corrupt, sometimes a most repulsive rule of conduct and practice, has often prevailed in whole tribes and families of mankind. Again, warm and earnest debates on the foundations of morality, and its primary laws, were transmitted from the clearest and keenest intellects of Greece to those of Rome. They remained still under ceaseless discussion for five centuries, until Christianity replaced them by other questions of a still deeper kind.

At the revival of learning after the middle ages, the flames of these ethical controversies, which had smouldered so long, broke out anew. And now, for more than three centuries, moralists and divines have rivalled the ancients in the zeal with which they have espoused conflicting theories. The strife, almost as keen and eager as in the days of Zeno and Epicurus, has lasted even to the present hour.

How is it, again, that when Physics have advanced with such giant strides, the students of morals, by the confession of some of its ablest writers, should hardly change their ground? Are they not still renewing the same questions, which occupied youthful disputes in the days of Socrates and Plato, more than two thousand years ago?

"There is no study," says Professor Grote, "more universal than Moral Philosophy. And yet, as a science, it cannot be said to have a high reputation at present in our own country. Nobody expects to learn much from what professes to be Moral Philosophy, or seems to think that much can come of it. This carelessness arises from a sort of notion that it is very likely to be mere words, or else a sort of quackery, very likely not to take hold of human nature, but to rest in useless generalities... Others are jealous of it, on account of its supposed tendency to level, regulate, and square human character, destroying its nativeness and variety; while those who are disposed to levelling and regulation are not in general interested in human philosophy of any kind."

These prejudices against the study, from the moral diseases and confusions of the world, the endless disputes of moralists, and the seeming absence of real progress, gain double strength from causes peculiar to these times. The wide spread of a school of thought, which denounces Theology as an impossible science, casts its dark shadow over Ethics also. It tends to degrade it into a secondary branch of physiology, dependent for its very being on the growth and progress of anatomical science. The advance of Physics in all its branches, and the many inventions with which it has enriched and adorned human life, bring out into greater prominence the seeming absence of similar progress in what claims to be a higher and nobler field of thought. The restless activity of a bustling age spreads around men a close and stifling atmosphere, like the fog and smoke which settle down on our crowded cities. High and lofty truths from the upper regions of thought seem often to strive, almost in vain, to penetrate this thick gloom. There can be no wonder that, in such an age, ethical science should often be exposed to silent neglect or open scorn.

In a Christian church and nation it might naturally be supposed that the clearer light of moral truth, found in the pages of a Divine message, would more than compensate for other causes of neglect and decay; and would secure for morals their due place in the intellectual culture of the age. And this is doubtless true within certain limits. The Decalogue and the Sermon on the Mount, with their popular expositions, have kept before the eyes of our people a standard which heathen nations never attained, and have secured a very wide acceptance for the simplest and plainest elements of Christian morality. But even this great gain, through human infirmity, has not been free from some attendant loss. The higher doctrines or deeper messages of Christian faith have not seldom been so abused as to deaden the general interest in the great lessons of Christian and revealed morality. Nay, even the fact that great moral precepts have been plainly revealed, and have been received in name by every Christian, may turn to a hindrance of all activity of moral thought. It may quench, instead of quickening, the thirst for a fuller knowledge of the grounds on which those precepts are based, of their various harmonies, their countless developments, and innumerable applications.

There is still another way in which the evidence of moral truth may have been clouded and obscured by its connection with the cause of Christian orthodoxy. Some of our ablest Divines, in maintaining revealed religion against assailants, have insisted strongly on the duty, in matters obscure and uncertain, of resting satisfied to act on probabilities alone. Thus Butler begins his Analogy with a laboured argument, to show that even a slight excess of evidence, in questions highly uncertain, may determine practically what course it is *right* to pursue. But however true in the abstract it may be, such a line of defence must needs exercise a depressing influence on minds really athirst for truth. It is very apt to give them the impression, that the being of God, the person and work of Christ, the atonement, the resurrection, and the life to come, are looked upon as having only a slight excess of evidence, if any, over opposite falsehoods. They may thus be tempted to infer that the scales of faith and no-faith tremble in an almost even balance, so that confident faith is only just one degree less foolish than confident unbelief.

Such an opinion must act like a palsy on all the nobler instincts of the Christian heart. It must surely have been a firm, deep, and hearty conviction, far unlike this timorous, hesitating, moonlight faith, by which the Apostles turned the world upside down, and overcame the might of the heathen empire, and martyrs poured out their very life-blood in defence of the truth.

This cautious, defensive line of thought, which Butler and other writers have employed in their advocacy of revealed religion, has naturally affected the kindred subject of morals. They insist strongly on the practical obligation of being guided by the slightest excess of seeming probability. The want of fuller evidence is said to afford scope and opportunity for a virtuous exercise of the understanding, and to dispose the mind to rest satisfied with any evidence that is real.

In such remarks there is a dangerous tendency to canonize doubt and uncertainty, and consecrate dimness of moral vision, as if they formed a useful moral discipline for Christian men. To seek and long for clear light in moral inquiries has sometimes been even discountenanced and condemned, as the mark of a sceptical spirit. But this is a dangerous inversion of the real truth. Such a desire, there can be no doubt, needs to be tempered with modesty and patience. There ought to be a willingness to be guided by moonlight, or even by starlight, until a clearer day-break shall arise. But still "light is sweet, and a pleasant thing it is for the eyes to behold the sun." The desire in itself is the healthy, inseparable instinct of an earnest mind and a Christian heart. It must be owls and bats of the moral world, and not the sons of light, who find it easy and natural, in the high matters of faith and conscience, to rest content with the dim twilight alone.

Let us now inquire whether these four difficulties, which seem to beset the approaches of Moral Science, will not disappear on a closer view.

First of all, the objection from the low standards of moral feeling and practice in savage tribes, from the revolting usages which have sometimes prevailed, and the great amount of vice and immorality even in Christian lands, can have little or no real weight with thoughtful men. No sensible person, certainly no Christian believer, can suppose for a moment that a perfect science of morals can issue forth, like Minerva from the head of Jupiter, complete, full-grown, and fully armed, from the mind and heart of the ignorant, the sensual, and the profane.

Every science, however sure in its principles, or fertile in its results, needs first of all to be learned. To learn requires not only some natural capacity, but attention, seriousness, and diligence. Ethics are a law of duty. How can it be learned by those who are reckless of duty, and care chiefly or only for animal pleasure? They are a light shining from above. How can it be seen by those whose eyes, like those of Mammon, are "always downward bent" upon the pavement below?

In the field of moral inquiry more is required of the willing learner than in the walks of physical research. Moral, as well as intellectual conditions, need here to be fulfilled. There must be a serious desire to learn and know the truth, however humbling to our pride, and however unwelcome to our indolence its lessons may be. Yet even in Physics, and in an age mentally so active as our own, how small is the number of those who obtain real scientific insight, compared with those who receive with implicit confidence the conclusions in vogue, from time to time, among scientific men. Let some feud arise amongst the known students and leaders in any science, and nearly all the seeming knowledge of these multitudes, being nothing more than opinions taken on trust from others, would quickly disappear.

Moral Truth, to be received and held firmly, needs an upward eye, and an open heart. Such is the voice even of natural reason alone. But Christianity lays a double stress on these necessary and indispensable conditions of all moral insight. "If any one be willing to do His will, he shall know of the doctrine, whether it be of God." "If thine eye be single, thy whole body," i.e. thy whole intellectual being, "shall be full of light." Once let us read the history of the moral aberrations in the world by the light of these true sayings of Scripture, and neither the parricide, infanticide, and cannibalism, of heathen tribes, nor the licentious vices, the drunken intemperance, which defile and dishonour Christian nations, will lead us to question the certain attainableness of moral truth. The proverb of Solomon will supply a brief and full answer to such causeless doubts: "Evil men understand not judgment, but they that seek the Lord understand all things." Humility, joined with earnestness and diligence, is the portress which alone can unlock the gates, and open our way to the temple of Moral Science.

The disputes and controversies among moralists themselves are a far more specious objection to the claim of Ethics to be deemed a genuine science. How can certain truth be attainable, where there seems to be neither harvest-field nor vineyard of ingathering, but only a battle-field of ceaseless debate?

The triangular duel of the Academician, the Stoic, and the Epicurean, has lasted for twenty-two centuries. Slightly varied in form, it continues to the present hour. The morality of sentiment stands opposed to the moralities of reason. A morality, apart from consequences, the categorical imperative of pure reason, fights with a morality reasoned out from consequences and results alone. How shall the disciples attain certain truth, when the leaders of thought, from age to age, seem to be in hopeless discord? It is the saying of Professor Grote, that "when we come to morals and ethics, the dust and smoke" common to other parts of philosophy "become tenfold worse." How should we look for peaceful certainties amidst the confused war-cries of a battle-field? *Silent inter arma leges.* The reproaches of unbelief against the various sects and divisions of Christianity seem applicable, with equal force, to impeach and annul the claim of Ethics to be a real science.

The answer to this objection, with a little patient thought, is not hard to discover. In every science there are some first principles, which are the starting-point or the foundation of all its other truths. These later truths form the pathway to which it leads, the building reared on that foundation. Now in every case these first principles, because they are the first, border on mystery, and lose themselves in the unknown.

"On what does the earth rest? On the back of a huge elephant. On what does the elephant rest? On the back of a huge tortoise. On what does the tortoise rest? Ah! that, said the Indian, I cannot tell." This Eastern tale or parable may be applied to the foundations of every science. A well-known and able writer has laboured to show, in the opening of one of his chief works, that the Ultimate Religious Ideas and the Ultimate Scientific Ideas are alike "unthinkable." By what strange legerdemain of thought the conclusion is reached, that Religion alone is mystery and nescience, and Physical Science a field of knowledge and progress, it is hard to understand. The difficulty is plainly common to both subjects, the natural on one side, the moral and religious on the other. The solution, so far as a solution is possible here, must be exactly the same for both. The striking words of Hooker apply here in all their force and beauty. "The goodliness of houses, the stateliness of trees, when we behold them, delighteth the eye. But that foundation, which beareth up the one,—that root, which ministereth life and nourishment to the other, is in the bosom of the earth concealed. And if at any time it be needful to search them out, yet is the search more profitable than pleasant, both to them that undertake it, and also to the lookers on."

Geometry has its half-solved or unsolved and perplexing problems, no less than Ethical Science or Religious Faith. It deals with the relations and properties of space. But what is space? Is it a substance, an accident, or a relation? Is it fixed and absolute, or relative only? Is it finite or infinite? Is Infinite Space a real something, or a mere expression for the "imbecility of human thought"? Can there be empty space, or must there be a plenum? Is it infinitely divisible? Can there be a surface without depth, a line without breadth, a point without length, breadth, or thickness? Velocity is the speed or rate of motion, or space divided by time. How can this vary from moment to moment, when both the space and the time, for any one rate of motion, wholly vanish and disappear? These questions, and a host of the same kind, furnish large materials for controversy and metaphysical debate. Had geometers forborne their labours till a perfect and clear solution of all these had been found, the science would be still unborn. Pythagoras would never have slain a hecatomb for his immortal discovery, the *Elements* of Euclid never have been written; the tomb of Archimedes, re-discovered by Cicero with natural triumph, would never have borne the trophies of his skill, the cone, the cylinder, and the sphere. The later triumphs of modern analysis would have been still more impossible. But they laboured on, and built still on the old foundations. They did not pause to grope beneath them in that region of mystery, on which all first principles must repose. For there our finite reason strives in vain to search out the unsearchable, or to attain that full insight and perfect knowledge, which the Omniscient God, the Only Wise, seems to reserve for Himself alone.

The course pursued in moral inquiries has not been altogether the same. There have been repeated attempts, it is true, and never wholly fruitless, to develop the consequences of the first principles of Ethics, and unfold them in systematic harmony. Dr Whewell's *Elements* are one example of this kind, well worthy of high praise and patient study. But, in general, more pains have been taken in re-examining the grounds of Ethics, or in exploring its debated border lands, than in tilling the soil, and reaping peaceful harvests. Or else systems have been reared on a doubtful basis, maintained by some, and by others stoutly denied. Thus an air of instability has often been given to the whole structure. Systematic Moralists are thus joined at various points in their progress, like troops in full march by stragglers, by many who dispute or deny some of their premises; and who, in the words of Jeremy Taylor, are "sure of the thing, even when they are by no means sure of the argument." But conclusions, however true and sound, when reached independently of the premises, or even in spite of their rejection, are wanting in all the main features of exact, well defined scientific truth.

But a third objection to the claim of Ethics, even more serious and vital than the discords and disputes of moralists, has been drawn from their supposed barrenness, and entire want of real progress. The contrast between the stationary aspect of moral, and the progressive aspect of intellectual truths, is said to be startling. "All the great moral systems which have exercised much influence have been fundamentally the same. All the great intellectual systems have been fundamentally different."

These statements of Mr Buckle, on the face of them, are excessive and untrue. The teaching of the Stoics differed widely from that of the Epicureans in ancient times, and both of them diverge widely from the ethics of the gospel. The contrast between Hobbes and Mandeville on the one side, and Cudworth and Clarke on the other, is not less in modern days. On the other hand, close resemblances may be found between some of the latest novelties of modern scepticism, with its pantheism, fatalism, and nescience, and the oldest forms of Eastern speculation. Still, when all abatements are made, there is a contrast not to be denied, between the rapid progress of Physics in these two last centuries, and the seeming want of growth in the field of ethical inquiry. What answer can be given to this reproach, while claiming for Ethics to be the highest and noblest part of all purely human science?

Now scientific progress may be of four different kinds. These may be styled briefly, ascensive, expansive, descensive, and diffusive. A science may climb higher towards those simple laws which rule over all its complex phenomena. It may range over a wider landscape, by unfolding these laws into a rich and large variety of secondary axioms. It may stoop down, to apply its discoveries more frequently and largely to the uses of daily life. It may gather around it a wider and wider circle of disciples, and may thus spread its light further and further in successive generations.

To begin with the last. The diffusion of Ethical Truth has been greater, not less, than that of modern Physics. Its first maxims, indeed, are so simple as to be easily mistaken for truisms, and excite the contempt of those who cultivate the intellect and neglect the heart. They may thus be decried as too commonplace to form the basis of any science worthy of the name. So to eyes accustomed only to gaslights and chandeliers, a fixed star at midnight may seem a worthless thing. It has always been there in the sky for ages, and its light is very small and feeble. And yet this simplicity of moral maxims favours their wide diffusion, and helps to secure their spread throughout the family of mankind. "There is no speech nor language, where their voices are not heard." The great law of truth, the voice which condemns treachery and falsehood, appeals to millions who never heard of, much less received, the Copernican and Newtonian theories. The law of kindness and goodwill has vibrated through myriad hearts, which never dreamed of the undulations of light, and to whom spectrum-analysis and polarization are mysteries unheard of and unknown. The law of self-sacrifice for the good of others, whether presented in word only, or embodied in the life and death of martyrs, or enshrined in the deepest mystery of our faith, has touched and stirred the deepest waves of thought in countless hearts, to whom nearly all the terms of modern science are sounds without a meaning.

The moral science of the past may perhaps be elementary in the extreme. But even in its weakness, and amidst its ceaseless struggle with prevailing vice and passion, it has exerted a far deeper power over millions of hearts and lives, than Physics have attained in the hour of their noblest triumphs, or can ever hope to attain.

Let us next compare Physics with Ethics in their descensive progress, or their practical application to the uses of human life. Natural Philosophy, since the days of Newton, has doubtless enriched society with a great multitude of new inventions. It has supplied comforts to the artisan and cottager, as well as multiplied luxuries and triumphs of art for the homes of wealth and the palaces of kings. But this kind of progress is no less real, no less extensive, in Christian Morals than in Physical Science. The virtuous man, the sincere and upright Christian, finds occasion every day and every hour to practise anew, and unfold in some fresh application, the great and simple precepts of Christian morality. The liberal man will continually be devising liberal things. The virtuous woman "opens her lips with wisdom, and in her tongue is the law of kindness." A pure fountain will hourly be sending forth pleasant streams. The quiet, ceaseless activity of great moral truths, once received into the understanding and heart of Christian men, may easily be overlooked by busy workers on the ant-hill of science, or else may be forgotten in the glare of some brilliant invention or new discovery. But it is not on that account the less real and important. In the words of Cowper,

> Stillest streams
> Oft water fairest meadows, and the bird
> That flutters least is longest on the wing.

I come now to that double progress of science, in ascent and in expansion, where Physics may seem at first sight to have a real superiority. Even here, I think, the claim will appear groundless on a closer view.

"There is nothing in the world," Mr Buckle says, "which has undergone so little change as the great dogmas of which moral systems are composed." On the other hand, those highest laws, which embody the clearest results of modern physical discovery, the law of attraction, the undulations of light, the correlation of force in various forms, have been reached by a slow and painful ascent in the course of ages. When once attained, they work a mighty revolution in our knowledge of the material world, and in our power to control its manifold changes.

Is this a proof that Ethics are lower than Physics in the scale of science, and moral truths more uncertain and inferior in practical value? Far from it. What is cheap and procured without labour is apt indeed to be despised. But even in the natural world, the air, the sunshine and light of heaven, the rains and dews that fertilize the earth, are gifts freely bestowed on men, and require no human labour. If the Divine wisdom has secured to all men, whether by the voice of conscience, or by supernatural revelation, direct access to the vital elements of moral truth, and has left the answering laws or principles of the outward world to reward the study of the philosopher, and to be explored by human toil, ought the free gift, on this account, to be despised, or consigned with contempt to the list of useless and barren truisms?

The law of gravitation is a noble triumph of human sagacity and persevering labour. After long ages of preparation, by Ptolemy, Copernicus, and Kepler, the profound study of Newton deciphered the handwriting of that Divine law, and expounded it clearly for the admiration of mankind. But the great law of duty—"Thou shalt love thy neighbour as thyself," however early or clearly revealed, is higher and nobler still. We may apply to it the verses of Sophocles, not in disparagement, but in praise;—

THE CERTAINTY OF MORAL TRUTH. 43

Οὐ γάρ τι νῦν τε κἀχθὲς, ἀλλ' ἀεί ποτε
Ζῇ ταῦτα, κοὐδεὶς οἶδεν ἐξ ὅτου 'φάνη.

What, then, is the main feature of that progress of astronomy, during the two last centuries, which is the chief boast of Physical Philosophy, and holds an undisputed primacy among its many signs of vitality and growing power? It is the unchanged recognition of one great law, and increasing faith in its perfect truth, amidst the immense complexity, the mathematical difficulties, and the ever-enlarging variety, of its results and applications. It forms thus a sublime example of permanence amidst ceaseless progress.

Other branches of Physics have not reached this goal, and their ultimate laws still elude research. There is change and uncertainty in the first principles themselves. The hypothesis of emission, even though sustained by Newton's authority, retires and gives place to a theory of undulations. This theory, again, in its first form of direct, is replaced by another of transverse vibrations. Caloric, the fictitious fluid of heat, disappears, and atomic motion comes in its stead. The old elements, as elements, pass away. Phlogiston is born, and has an early death. Forty or fifty new elements occupy the foreground. But they hold their place by a precarious tenure, till chemistry takes a further stride. The simplest elements, oxygen and hydrogen, now represent in their names the ghost of a departed theory. Magnetic and electric fluids come like shadows, and like shadows they depart. Here science is still only on the steep and slippery mountain side. Though it may wield its ice-axe with steady skill, it mounts slowly and with labour. And not seldom it slips backward, when some hasty, seductive hypotheses, like a crevasse thinly covered with new-fallen snow, betrays its insecure foothold. But astronomy, since the days of Newton, seems to stand firm on the mountain-top, and gazes calmly on an ever-widening range of peaceful scientific victories.

Now whence arises this manifest superiority over other branches of physical study? The reason is plain. Astronomy has attained a comprehensive simplicity in its primary law, and fountain truth, such as Ethics, with the aid of a Divine message, had attained long ages ago. For what is the second great commandment? It is a higher law than that of universal gravitation, binding together the whole universe of moral agents. And it has been revealed, more dimly by natural conscience, but in plain and express terms by the great and divine Author of man's moral being.

The simplicity and terseness, then, of the first principles of Morals, is not, as the modern historian falsely conceives, their shame and weakness. It forms rather, as the great tragedian of Athens more truly saw, their strength and their glory. The roads and pathways of earth change continually with the revolutions of human society. Mule-tracks are replaced by turnpikes or Roman causeys, and these, in course of time, by iron railroads. But in the blue sky there are no such fashions of the age. The courses of the stars are silent and unchangeable.

Moral truths are the stars of the intellectual world. They differ from those lights, useful and yet earthly, which men dig up from the mine, distil in their retorts, conduct through pipes, and transmit to burners, all of them the laborious products of human skill. They shine by their own native light. They look down upon us calmly from above, unless when fogs and mists of earth intercept their brilliance. They are waymarks for the benighted, and also dim heralds of the coming day-break. "Men may come, and men may go,"—their mines may be exhausted, their pipes burst, their candles flicker and waste away,— but these stars are immortal, and shine on for ever.

But has not experience at least proved an immense contrast between moral maxims and the laws of Physics in their power of expansion and development? The first are stationary and barren. Nothing has been added to them, it is said, by the sermons, homilies, and text-books of a thousand moralists and divines. The others are fertile, active, and progressive, and receive from year to year a more full and large development.

The contrast, though greatly exaggerated in the remarks to which I refer, has doubtless a partial truth. The development of the great truths of Ethics, in modern times, has been far less apparent to common observers than the growth and progress of Physical Science. A fuller knowledge of the best moralists and divines than physical students, and especially positive philosophers, usually care to attain, will mitigate the sharpness of the alleged contrast, but cannot wholly remove it. When, however, we look below the surface, the seeming defect will be found to have its source, partly in the moral evil which interferes with and clogs all ethical study, but partly in the superiority and higher dignity of Moral Science.

In Physics the development of theory, and its application to the practical uses of life, are to a great extent independent of each other. The theory of magnetism might still advance to new discoveries, though timorous navigators were to abandon the use of the compass, and to creep, as in early times, timidly along the shore. Art may grow by happy accidents, and the increase of practical cleverness, while the answering science remains stationary. Science may take immense strides in advance, before its discoveries begin to be applied in new inventions, and yield fruit in the service of daily life.

In Morals the case is wholly different. Its true aim, as Aristotle observes, is "not that we may know what virtue is, but that we may become good men; for else it would be useless." It is thus essentially a practical science. It teaches men what they ought to be, what they ought to do. Its theory, when wholly divorced from practice, is not only useless, but even mischievous. It aggravates and redoubles guilt. "To him that knoweth to do good, and doeth it not, to him it is sin."

Thus in Moral Science theory and practice are not capable, as they are in lower sciences, of a wholly separate development. Corrupt, selfish, vicious practice, must cloud the eye of the soul, and impair or destroy the faculty of moral discernment. Knowledge of the truth cannot grow, unless the truth already known be received into the heart, and applied in the life. In Physics speculation is like an engine with no train attached, which can thus move more freely, and advance at a swifter pace. In Moral Science it is so closely linked with the whole moral being, that the mind, heart, and life, the conscience, intellect, and affections, must advance or recede, must rise or fall together. In a world, then, where moral evil plainly abounds, it is no wonder that Ethics, even as a science, should make only a slow, intermittent, and faltering progress. Its standard, raised higher here and there for a moment, must fall again whenever luxury and selfishness abound. Progress made by the calm and patient meditations, or the heroic virtues of a few, may be lost, to all outward appearance, by the empty folly or fierce and angry passions of the many. The fires of some scorching judgment may be needed from time to time, before conscience can burst the chains of reigning sensuality, shake itself from the dust, escape from its cave of shadows, stand erect in the freedom of a new life, and gaze with clear vision and open eye on the grand realities of the spiritual world.

There is no reason, then, from the limited progress of Ethical Science in time past, to depreciate its worth, or to doubt its capacity for vast and large expansion in days more favourable to its development. In the sacred words of its Lord and Master, "its hour is not yet come." It has now to struggle onward and upward under the burden of a host of sluggish or perverse disciples, ready to turn aside into seductive by-paths, or to sink lower and lower in vice and folly. These load with a heavy weight, and weaken with a stifling atmosphere of moral impurity, even those willing learners who lend themselves to nobler impulses, and strive to rise. The allegory in *Comus* well describes, side by side with its real worth and divine effects, the contempt with which it has thus come too often to be regarded by some physical students:

> The leaf was darkish, and had prickles on it,
> But in another country, as he said,
> Bore a bright, golden flower, but not in this soil:
> Unknown, and like esteemed, and the dull swain
> Treads on it daily with his clouted shoon;
> And yet more medicinal is it than that moly
> Which Hermes once to wise Ulysses gave.

In a happier age, assured to us by Divine promise, that bright, golden flower will appear. Or rather, in the sacred words of Scripture, it will become "a tree of life, yielding its fruit every month," and its leaves will be for the healing of the nations.

I must reserve for another Lecture the fourth and last hindrance to the claims of Moral Science in the doctrine of probability, and some direct enforcement of the certainty of moral truths. But before I close I must add a few words on the vast importance of a full confidence in their authority, reality, and unchangeableness, to meet and overcome the special dangers of the time in which we live.

Amidst the busy activity of commerce, the excitement of political strife and change, the progress of physical inquiry, and all the feverish, impatient movements of modern thought, careful observers must observe one symptom full of danger. Multitudes, in every Christian land, seem to be more and more uncertain on every subject that rises above man's animal wants and desires, and which touches on the unseen and eternal. The Christian Church, the domain of Theology, is vexed with endless controversy and division. Rival bodies of Christians seem often more jealous of each other than zealous for common truth. Political life is in danger of sinking into a mere strife of parties, ranged under doubtful banners, ill deciphered and guessed at by their own followers,—a battle without generals and without discipline. One tide-wave of popular impulse follows another in swift succession. All old landmarks are ready to disappear beneath the shifting flood. But where shall certainty be found? Where can the toil-worn spirit of man find solid footing, and attain a sabbath of rest? He can never be content to find it in the properties of numbers, in squares, triangles, and polygons alone. A secret instinct whispers to him, even when most debased by folly, that his true portion must be sought, and can only be found, in a higher and nobler field. Tell him that morals and religious faith are a Hyperborean region, buried in fog and mist and perpetual darkness, where endless uncertainty reigns, and doubt, sometimes honest, but more frequently dishonest, is the highest attainment, and some slight excess of the chances in its favour the only reason for a virtuous life, and what results will follow? Such a view will palsy the arm of the Christian, when uplifted for noble action, and strike an icy coldness to his heart.

Can we, who inherit the transmitted wealth of thought of so many ages, not from heathen philosophers alone, but from prophets and apostles, and the great lights of the Church of Christ in all past ages, have fallen back, in these last times, into disgraceful contentment with more than heathen darkness? Are we ready to accept, as the teaching of advanced philosophy, a creed of pure nihilism, that Theology is a dream, religion pure nescience, covered thinly with a crust of decaying dogmas and idle fancies,—conscience a strange function of nervous tissues, or at best the acquired instinct of slaves, who crouch sullenly under the lash of masters stronger or more numerous than themselves? Then indeed our boasted light can only be paralleled by the state of Elymas the stricken sorcerer, when there fell upon him "mist and darkness," and he groped about for some one to lead him by the hand.

Such a view of Moral Science, as a field of mere uncertainties, is gloomy in itself, and disheartening and melancholy in the results to which it must lead. Our duty as Christians, and even as students in this great university, doubly adorned by the greatest lights in Physics and Theology, is to hold firmly a more hopeful and far nobler view.

Ethics is the Science of Ideal Humanity. It deals with that lofty standard of the good, the right, the true, the noble, towards which all of us, from the least to the greatest, are bound continually to aspire. And this can be no land of mist and shadow. Even in times of heathen darkness it had a starlight of its own. The Gentiles, who had no direct revelation, were a law unto themselves. And now, in the times of the Gospel, it has a light as of the daybreak, which shines more and more, in honest and loving hearts, unto the perfect day. Its lowest regions, indeed, must partake largely of the uncertainties and complexities of that human life with which it has to deal, with its manifold conditions of bodily faculties, mental powers, diverse tastes, and different social relations. But above and beyond these there is a higher region, where it rises above the complex and ever varying applications of its own laws, and deals with the inward motives, affections, and desires alone. And here it breathes a purer air, it eats angels' food, and walks in the light of heaven. The firmness of its deep foundations below is only rivalled by the pureness and clearness of the light that shines upon its higher summits from above.

To such an Ethical Science, so firmly rooted in essential truth, and rising to such a lofty elevation above the mists and clouds of sceptical uncertainty, which vex and obscure the minds of busy multitudes in this present age, we may apply the description of one who is supposed to have practised its high lessons, and to have taught them to his simple flock, the pastor of the *Deserted Village:*

> As some tall cliff that lifts its awful form,
> Swells from the vale, and midway meets the storm;
> Though round its breast the rolling clouds are spread,
> Eternal sunshine settles on its head!

LECTURE III.

THE CERTAINTY OF MORAL TRUTH.

EXPERIENCE AND REASON.

THE doctrine that in all moral subjects probability is the highest attainment, is a part of the wider theory of the ancient Academy, and has been held by many moralists in later times. Aristotle remarks that "it is the duty of one well instructed to seek for accuracy in every kind of subject, only so far as its nature allows;" and that "it is much the same error to be content with mere persuasion from a mathematician, and to require demonstrations from the rhetorician." "Every discourse," he adds further, "about things to be done must be in outline, and not exact." He seems here to contrast geometry with ethics, and to impute inexactness and uncertainty to the latter, but to practical or applied ethics alone. The later Academy held the principle more widely, that in all things probability was the utmost limit of the human mind, from the fallibility of the senses, and the difficulties and seeming contradictions in every field of thought.

Bp. Butler, again, is led to insist often on the responsibility which results, in moral and religious subjects, from even probable evidence and imperfect half-knowledge.

But this line of thought is often carried so far, as at least to verge on a dangerous extreme. The doubtfulness of the evidence, in such cases, is sometimes assigned to Divine appointment, in order to vary and enlarge the field of moral discipline and probation. The strict discharge of duty, he says, with less sensible evidence, does imply a better character than the same diligence in the discharge of it on more sensible evidence. "And men are bound," he observes, "to act on what appears to them, to the best of their judgment, to be for their happiness, as really as on what they certainly know to be so." It is plain, however, that mere guesses and probabilities, even when they involve a practical duty to act upon the more likely of two alternatives, can never form the materials of a genuine science.

"Physical and Moral Studies have a common ground; science, and not probability, is the aim of both." Such is one of the latest words of Professor Maurice in his Lectures on Social Morality. The principle thus affirmed is one of high importance. It might almost claim to be the foundation of Morals as a Science. I do not think that it is to be viewed as the consequence and result of worship. It is rather the indispensable condition that worship may be a reasonable service, and not sink into a blind superstition. It is a conviction by no means limited to those writers who lean to a mystic or transcendental school of thought. The judgment of Locke in its favour has double force; since it might have seemed natural for him, from the general tone of his famous treatise, to rest content with a lower view. He writes as follows:—

" The ideas of quantity are not those which alone are capable of demonstration and knowledge. Other, and perhaps more useful parts of contemplation would afford us certainty, if vices, passions, and domineering interests did not oppose or menace the endeavour."

"The idea of a Supreme Being, infinite in power, wisdom, and goodness, whose workmanship we are, and on whom we depend; and the idea of ourselves, as understanding, rational beings, being such as are clear in us, would, if duly considered, afford such foundations of our duty, and rules of action, as might place Morality among the sciences capable of demonstration. I doubt not but, from self-evident propositions, by necessary consequences, as incontestable as those of Mathematics, the measures of right and wrong might be made out, to any one that will apply himself with the same attention and indifferency to the one, as he does to the other, of these sciences."

"Two things," he continues, "have made moral ideas thought incapable of demonstration, their complexness, and their want of sensible representation." "Confident I am, that if men would, in the same method, and with the same indifferency, search after Moral as they do after Mathematical truths, they would find them have a stronger connection one with another, and a more necessary consequence, and to come nearer perfect demonstration, than is commonly imagined. But much of this is not to be expected, while the desire of esteem, riches, or power makes men espouse the opinions in fashion, and then seek for arguments either to make good their beauty, or to cover their deformity. For nothing is so beautiful to the eye as truth is to the mind; nothing so deformed, and irreconcilable to the understanding, as a lie."

This view of Locke, Dr Whewell, and Professor Maurice is exposed, no doubt, to plausible objections, and beset with seeming perplexities, but still it must commend itself spontaneously to thoughtful and generous minds.

"There is no reason," I have said, "in the nature of its first principles, why the evidence of Morals should be placed on a lower footing, or regarded as less demonstrative, than that of Geometry....It is as easy to conceive the laws of space inverted as those of conscience, and to grant that the properties of the circle might change, as that treachery should become a virtue, and compassion a crime....The spirit of man has its own self evidence, conscience its own definitions. And these are not cold and silent, as in Geometry, but vocal with strong emotion, and endued with a living power. Its truths are in their own nature as prolific. The opening for deduction is the same. Why should it not lead to certainties as sure, and conclusions as various and manifold in this higher and nobler field?"

Here, at the outset, two admissions have to be made. And first, if the doctrine of consequences is the only test of right and wrong, and such actions only are to be esteemed morally good, as can be proved in their result to secure "the greatest happiness of the greatest number," Morals can never be raised to the level of a genuine science. It must reduce itself to a loose heap of uncertain probabilities alone. A strictly utilitarian science of Morals would require for its construction nothing short of Omniscience. To compare the moral character of two alternatives, we should then have to construct two future worlds of thought and action, composed of moral agents living and unborn, reaching on through the most distant ages, and then to decide to which of these possible future worlds the greater sum of collective happiness would belong. The problem is plainly impossible in itself for human knowledge. But the impossibility is multiplied a hundred fold. For we should have to treat all the future actions and feelings of these countless moral agents as fixed terms, except so far as they depend on this one alternative choice of one agent alone. Now, in reality, each of them is unfixed and variable in the highest degree. The attempt to found anything worthy the name of science on such a basis must be a deception and a failure. The seeming solution can only be gained by the arbitrary selection of a few elements of an impossible problem, and omitting a thousand others, equally required for a genuine solution. Loose guesses, in which nearly all the data are wanting, must be confounded with strict demonstration, and a power of just and true forecast be claimed for human ignorance, which belongs to the All-seeing Wisdom alone.

Again, the claim of a strictly scientific character, however justly made on behalf of pure Ethics, or the great rules and principles of personal, social, and religious duty, cannot be extended without serious error to its practical applications. It may be a clear and certain truth that we are bound to seek the good of our neighbour. It must be a complex and doubtful inference, that we shall best promote his welfare by this or that particular act or mode of action. There is a time to give, and a time to refrain from giving; a time to speak, and a time to keep silence.

Moral principles, however firm and sure in themselves, can only be reduced to practice by joining them with the lessons of a wise expediency. And this expediency cannot rise higher than the level of its source, which is the defective knowledge of human nature and all outward circumstances, and the imperfect foresight, which belongs to fallible men. This contrast between pure science, and the elements of uncertainty in its applications, is conspicuous in Arithmetic and Geometry. Actual enumeration in many cases, and actual measurement in all, is approximate only. It can be no matter for surprise, and no proof of inferiority, that the same remark belongs equally to the practical applications of Moral Science.

How far, then, within these needful limits, may we claim for Morals a strictly scientific character, closely resembling that which belongs to Arithmetic and Geometry, and Pure Mathematical Science?

Knowledge, in every subject, may be of two kinds. It includes the actual and the necessary, what simply is, and what must be. In the highest object of thought alone the two conceptions meet together, and melt into one. The great I AM, is at once the Supreme Reality, and the necessary, self-existing Being, He who is, and cannot but be. But in every other case, in all the objects of sense, and the varieties of created and dependent life, the two kinds of knowledge are distinct. They are the separable objects of observation and reasoning.

We observe things and persons, and thus know that they exist. But we are conscious, all the time, that it might have been otherwise, that they might not have existed, such as they are. We can conceive of others in their place, wholly different. On the other hand, we frame an hypothesis, and deduce from it by reasoning certain results. But we may still be uncertain whether the premises and their results alike have any home in the world of actual realities.

The separation of these two elements is never total. The progress of science consists mainly in successive attempts for their more complete reunion. Phenomena cannot even be registered without the aid of memory. Now memory implies some thinking agent, to whom its

THE CERTAINTY OF MORAL TRUTH. 57

successive acts belong. On the other side, the formation of an hypothesis also presupposes such a thinking agent, and the existence of some real objects, to supply needful materials for the exercise of imaginative power.

Experience and reason thus approach every subject of human thought from opposite sides. One directly observes what is, and might have been different. The other traces the necessary connections between things which either are, or might possibly be. Complete, full-grown science consists in their perfect reunion. Reason, as we approach this limit, becomes more and more practical; till at length its hypotheses are merged and lose themselves in the real laws of the universe. In this way they become the source of mastery over nature, by genuine foresight of all future change. Experience, again, travels upward, from transient phenomena to things and persons; and from things and persons, seen as capricious and uncertain in all their actings, to the double supremacy of physical laws, and moral, intelligent design; till the real and actual loses itself in the ideal. All transient phenomena are then seen to belong to one vast scheme of Providence, firmly bound together by laws, natural, moral, and spiritual, some of them discovered by man, others reserved in the depths of Omniscient Wisdom, but all alike leading upward, and meeting at the footstool of His eternal throne.

In Pure Science the elements borrowed from experience are of the simplest and most general kind. Hence they almost disappear from view, and our attention is fixed on the reasoning process alone. Here, then, probability has no place. The facts which need to be silently assumed are so clear and simple, that they may almost be called self-evident. The reasoning, drawn from few and simple premises, is demonstrative and exact. The opposite of its conclusions is discerned, not simply as untrue, but as an impossible contradiction. The sum of two odd numbers cannot be an odd number. The diagonals of a square cannot be unequal. Two right lines, which meet in one direction, cannot meet in the other also. You cannot, in showing kindness to others, show more kindness to each of them than to all the rest. You cannot inscribe a hexagon in a circle with sides greater or less than the radius. But with experimental knowledge the case is widely different. The evidence on which it rests may vary from the plainest and fullest to the most uncertain and obscure. Its materials are of unequal value. They include present, direct observation, our own memories of the past, in some cases full and clear, in others obscure and doubtful; and also the observation and the memory, clear or obscure, of other witnesses, who vary most widely in veracity, caution, and intelligence.

The Applied Sciences consist either of empirical laws and secondary axioms, derived immediately from observation and experiment; or else in the combination of pure science with experimental data, varying in their degrees of accuracy and completeness. They must therefore share, more or less, in the uncertainty which belongs to human experience. There are doubts which may arise directly from the nature of the materials, as when the astronomer has to put aside faulty sets of observations. But there are others which depend on their own scientific imperfection. They may contain empirical laws, derived from an induction too fragmentary and imperfect. They may strive to account for one class of known facts by inventing another and larger class, which may possibly be true, but for which there is no direct evidence whatever. Their axioms may often be drawn from popular notions. They will thus be vague, elastic, and variable in sense, and infect with uncertainty and vagueness the whole course of reasoning. The tendency of the homogeneous to become heterogeneous, and the persistency of force, when force is made to mean half a dozen different things, are signal examples of vague, unscientific, deceptive substitutes for the certainties of science. Assumptions may be separately made, and reasonings founded on them, which do not agree, but really exclude each other. Or finally, their complexity, usually greater than in pure science, which abstracts the most needful elements in its problems, may be so great, as to place exact reasoning on them almost beyond the reach of a merely human intelligence.

There are two main reasons which may be offered, to throw doubt on the attainableness of moral truth, and to reduce its evidence to the class of uncertain probabilities alone. The first is drawn from its place in the scale of human thought, and the second, from the freedom and variety which seem implied in the actings of voluntary moral agents.

When we rise in the scale of being, from Nature to Man, from lifeless matter to plants, from plants to animals, and amongst living animals from the lowest to the highest, the complexity of the materials of science increases at every stage. The first principles themselves, so far as they have been discovered, are more complicated; and they shroud themselves more and more in a mysterious darkness, which it is hard for the students of science to explore. It might seem, then, at first sight, that Morals, the highest part of Human Science, must partake above all the rest in the uncertainties of observation and experience, and be removed the farthest from that character of certain, definite, demonstrable truth, which belongs fully to the Pure Sciences alone.

The first answer to this doubt may be found in the important fact that Morals are an ideal science. They propose to define, not what men have done, or may do, but what they ought to do. Truth is one, falsehood is manifold. The right in morals, as the name implies, has a close analogy with the straight line in geometry, in contrast with curves of single or double curvature. The properties of triangles and squares are far simpler and easier to investigate than those of circles, ellipses, and other figures of still higher orders and more complex definition.

Now this ideal character of Morals tends greatly to diminish the complexity of the problems of ethical science. The rules of right and wrong may be comparatively simple, when the question, how such and such persons are likely to act, is difficult and uncertain in the highest degree. The complex relations between the body and the soul, between man and the series of lower animals, the varieties of personal taste and disposition, the countless combinations of varying circumstances, may, in the higher regions of pure Ethics, be left almost wholly out of view. They belong essentially to the secondary results and applications of moral principles, and to these alone. There is thus a compensation, highly important, to that extreme complexity which might else be inferred from the place of Ethics in the scale of science.

There is also another principle to be kept in view, which conspires to the same result. In the ascending scale of science we rise, it is true, from the contemplation of simpler and lower powers to others of a more complex and higher kind. But we rise also from multiplicity towards unity, from a region of seeming confusion, and changes without visible purpose, to one of light, order, harmony and design. Chaos, or unformed, lifeless, multitudinous matter, comes first and lowest in the scale of being, and the well-ordered Cosmos stands higher. In the Cosmos itself, Man stands high above the brute creation; and highest in those characters which they do not share, in which he is contrasted with all the rest, and distinct from them, as created at first in the image of God.

All creatures, it has been said, strive to ascend, and ascend in their striving. And so far as this effort really exists, it includes a striving after unity, such as when a regiment of soldiers, parted in the smoke and confusion of a battle-field, seek to range themselves once more under the standard to which they belong. Man himself is one order, one genus, and one species, in contrast with the vast number of orders, genera, and species, by which he is surrounded in the animal world. But the Creator himself is One, in contrast to all the innumerable millions of mankind.

It follows naturally from this truth, that when we leave the simplest conceptions or abstractions of number and space, and mount higher in the scale of being, that complexity which renders scientific treatment almost impossible, and throws us back on mere experience and observation, is greatest in the middle stages of our progress. The mist, even in the intellectual world, often lies thickest on the sides of the mountain. When we climb still higher, there is a growing amplitude and grandeur in the wide firmament on which we gaze, and often in the earthward prospect, but we may also discern increasing signs and tokens of a Divine simplicity.

Astronomy, in the extent of space with which it deals, has a far wider range than Geology. And yet there is a grand simplicity and unity in the celestial motions, which make them far more accessible to scientific treatment than the waves of the ocean, or the solid strata of the earth. And so, when we rise to the highest part of men's nature, and from action to speech, from speech to thought, feeling, and desire, we reach at length that highest field of Human Science, where it borders on Theology. Here, even in his weakness, Man rises into communion with things unseen and divine. And here also tokens of a simplicity, adapted for the deductions of science, and difficult to discover midway in the long and steep ascent, meet us once more.

The great chain of thought and being seems thus to be firmly anchored, and fixed in scientific certainty, at both extremes. It is fastened firmly below to the truths of arithmetic and geometry and mathematical reason, which underlie the whole material creation. It is fastened above, no less firmly, to the throne of God. But its middle portion seems to hang more freely in space, and to oscillate amidst the endless obscurities and uncertainties of creature volition. It is only when, in the study of man, we leave behind us all lower elements, and confine our thoughts to the distinctive features of human reason and will, that the chaos of blind, ceaseless, purposeless change begins to recede from view, and the light of scientific certainty, growing out of a Divine unity, slowly dawns upon us once more.

It may be objected, further, to the doctrine of the fixedness and certainty of Moral Science, that it is opposed to the freedom of the human will, and would tend to mechanize and freeze down to a cold, bare, and heartless uniformity, the whole course of personal and social life. Is the law of right too complex to be capable of discovery? Then it must be practically worthless, and of no avail. Is it capable of discovery? Is there one line of action, which it prescribes, excluding every other as vicious, defective, and unholy? Must it not, then, condemn all who obey it to a dull monotony of action? Will not moral agents thus become only like material atoms, which obey the impulses of surrounding bodies, and have no power to deviate one hair's breadth from the path which inexorable laws have prescribed? Will not men thus be tempted to think of virtue as another name for tameness and dull monotony, and associate vice itself, breaking through these irksome rules, with the ideas of variety, life, activity and freedom?

If we accept the notion which Mr. Mill ascribes to most utilitarian thinkers, that morality deals exclusively with actions themselves, and not at all with the motives by which they are guided, and on which they depend, it will be hard indeed to rescue it from this serious indictment. It will then prescribe in every case some precise line of action, deduced from a calculation of consequences, and thus reduce all virtue to the level of a precise, inelastic, and cold routine.

This view involves, however, a great misconception of the real aim of Moral Science. Its true lessons seem to me very different. It lays down principles and motives, which ought to be the living fountains and sources whence our actions are to flow. It discloses, also, limits on the right hand and on the left, which they are bound not to transgress, and within which they ought to move. But when these conditions are fulfilled, it still leaves a wide and various range for the exercise of voluntary choice and human freedom. Its aim and purpose is not to annul and destroy, but to redeem from a fatal bondage, and apply to its noblest uses, that freedom of will and choice, on which its own existence, as a science, wholly depends. Its true type and keynote is found in the earliest Divine command revealed in the word of God; where, side by side with one special prohibition, there was expressly conceded a wide and large latitude for the free choice of the human will. "Of every tree that is in the garden thou mayest freely eat."

True morality is fixed and definite in its principles, and in the laws and limits of action it enjoins; but free, large, and various in its application to the complex economy of human life. It is not meant to extinguish, but to develop to the utmost, that genuine freedom on which its own authority depends. It is a law, but in the truest sense a law of liberty. It is firmly rooted in the soil below. But above it drinks in the free sunlight, ramifies into a thousand branches and branchlets, and effloresces freely amidst the play of the breeze, and under the light of heaven.

Two things seem chiefly required to vindicate the claim of Morals to be a genuine and certain science—simplicity and clearness in its axioms and first principles, and a capacity, in those axioms, for various and exact development. Let us briefly examine each of these points in succession.

The principle which lies at the basis of all geometry is the consciousness, gained through our senses, of the existence of an outward world. Moral Science rests upon a truth not less simple, and not less deep in its intuitive certainty—Man's inward consciousness of will, and of a power of choice, dependent on the inward activity of the mind and heart. It is a primary law of thought, one of the intuitions of the soul. "The mind is led by its very nature and constitution to perceive that there is an indelible distinction between good and evil, just as there is between truth and falsehood." "It embodies itself in the pronouns of every language, breathes in every desire of the soul, and lives both in the memories of the past, and the hopes of the future." It is the grand fundamental keynote, ever sounding anew in the daily voices of life, and the whole history of the world.

Pure Ethics deal immediately with the tempers, desires, and affections of men, viewed in the light of this one simple truth, that they are beings endowed with choice, reason, and will. Applied Ethics deal with the outward results of these same feelings, tempers, and desires, when carried out in speech and action, and ramified into union with all the physical conditions of man's complex being, and all the immense and manifold varieties of his social life. Thus Applied Ethics belong to mankind alone. But Pure Ethics must extend to every conceivable race of moral agents, or beings endowed with choice and reason, however various their other characters may be. It includes angels, as well as men. And if our fancy chooses to people distant planets with rational animals, widely differing from ourselves in shape, sense, and physical structure, the great laws of moral obligation must include these also. They do not depend on the specialities of man's animal nature, but on that gift of a reasonable will, which marks him out as higher than the whole brute creation.

Thus Pure Ethics, in the subject with which it deals, has a simplicity which rivals the science of number and of space. It deals with moral units alone. It leaves out of sight the secondary features of human nature. It looks on man simply as a moral agent, gifted with the power of voluntary choice, and thus capable of right and wrong desires, of good and evil actions. This view of his nature gives birth at once, by an instinct invincible in itself, and present even when most obscured, to a deep sense of moral obligation. Such a being is not under the constraint of a mere physical law, which he must obey. Yet he cannot be altogether lawless. His conscience owns a higher law, which he *may* break, but *ought* to obey.

The relations of such moral agents, the monads or atoms of the moral world, to the sentient beings around them, are in their own nature simple and uniform, like those of the points which form the triangle or the square. To the awakened conscience they are fixed, clear, and certain, like the truths of geometry. Only they are far higher in kind. Their voice speaks not only to the understanding, but to the heart. And hence there is no reason, from the nature of its first principles, why Pure Ethics should not form a clear and definite science.

But a science cannot be formed of a few axioms alone, unless these are capable of a real expansion and development. Are moral axioms, then, sterile in their own nature? Are they ice-bound as in an arctic frost, where life and growth can find no place? Such is the view, as we have seen, in the "History of Civilization." Nothing, its author tells us, has undergone so little change as moral axioms. They have been known for thousands of years, and moralists and divines, with all their labour, have not added one jot or tittle to their amount. Morality admits of no discoveries. It is cursed with perfect immobility and barrenness, while Physics pursue a triumphant course of never-ending discovery.

Now so far as a contrast may have arisen from the vicious habits of men, or their moral dulness or perverseness, it can have no weight to bar the claim of Ethics to be a genuine science. It must be shown the difference depends on the very nature of moral truths, and not on the backwardness and reluctance of men to receive them simply, to search them out honestly, and to explore patiently the results to which they lead. Is the contrast then real, or is its seat to be found in a diseased judgment alone?

Let us consider the great maxim of the Divine law—Thou shalt love thy neighbour as thyself. Its simplicity is extreme. But the law of gravitation is also very simple. "Every part of matter attracts every other with a force inversely as the square of their distance." And still, simple as it appears, the most profound geometers and analysts, for more than two hundred years, have tasked their powers to the utmost in tracing out and exploring its necessary consequences. The results of their labours are embodied in treatises and essays without end, in lunar, planetary, and cometary theories, in theories of tidal motion and the figure of the earth, in essays on variation of constants, nutation, precession, in the discovery of fresh terms of lunar and planetary correction, of growing minuteness, in the orbits, periods, and inclination of double stars, and in the triumphant addition of two new planets to the solar system. And they are still far from having reached the limit of possible discovery, or exhausting all its treasures of hidden truth.

Why should the Divine rule of duty, revealed in the Law and the Gospel, and confirmed by the voice of natural conscience, be less fertile and various in the deductions that flow from it, and the results to which it leads? It resolves itself at once into the double enquiry—Who is my neighbour?—How ought I to love myself? This latter enquiry divides itself next into the twofold obligation, to seek health of body, and health of mind, in intellectual and moral well-being.

The pursuit of bodily health includes many plain lessons of duty, in temperance, soberness, and chastity, active exertion and reasonable rest. But it also includes others less plain and self-evident, the duty of growing attention to sanitary laws, the due and proper use of medical skill and science, and watchfulness against the occasions, inlets, and provocatives, of dangerous disease. The duty has also many varieties and special forms, according to differences of occupation, and varieties of temperament, habit, and personal constitution. Thus it tends to absorb and incorporate all the successive discoveries of medical learning or practical experience. It condemns, as immoral and guilty, the careless ignorance, the blind improvidence and headstrong folly, which are like a partial suicide, and whereby so many not only endanger or shorten their own lives, but expose their fellows to the ravages of infectious and fatal disease.

The duty of self-love, when applied to the mind itself, opens a still wider field of thought. It parts at once into two main divisions. It includes the pursuit of all truth and light in the understanding, and of all pure and noble affections in the heart.

The pursuit of truth, again, is threefold, in an ascending climax of worthy objects of desire. It includes natural, moral, and spiritual truth. And each of these is vast and large in dimension and variety. The truths to be sought after, and eagerly prized and retained, become larger, nobler, and loftier, the higher we rise in the scale of being. Like the chambers in the temple of Ezekiel "there is an enlarging, and a winding still upward." The knowledge of nature is a wide field. The knowledge of Man, the free and moral agent, the lord of nature, is wider and nobler. And highest and noblest of all must be the knowledge of Man's Creator, of God the Only Wise, the Eternal Fountain of all wisdom, goodness, and love.

That province of self-love, which prescribes the culture of the affections, and the control of the desires, is no less wide and various. For these affections admit of being doubly classified, by their internal characters, and their outward objects. When each of these is once submitted to a distinct inquiry, the field of thought it discloses will grow wider and wider continually. It includes faith, hope, love, patience, meekness, zeal, courage, prudence, humility, nobleness of mind and heart. It views each and all of these, in their reference to all the varied relationships of human life, parent and child, husband and wife, brother and sister, master and servant, neighbour and friend, subject and sovereign. Everywhere, as we advance, new vistas of light and beauty will open before us. The beautiful words of Milton on the benefits and pleasures of education will apply still more fully to these researches of moral science:—" We will conduct you to a hill-side, laborious indeed in the first ascent; but afterwards so smooth, so green, so full of goodly prospect, and melodious sound on every side, that the harp of Orpheus was not more charming." And we may sum up the scope and compass of its teaching in the more sacred words of the Apostolic command: " Whatsoever things are true, whatsoever things are honest, whatsoever things are just, whatsoever things are pure, whatsoever things are lovely, whatsoever things are of good report—if there be any virtue, and if there be any praise, think on these things!"

But the other question—Who is my neighbour? implied in the same command, opens another wide and ample field for ethical study and meditation. For it leads us to reflect on the definition of moral neighbourship, and on all those special circumstances of opportunity, kindred, sex, age, and social position, which modify the general duty imposed by the great law itself, and diversify it into a thousand forms. It extends, in its widest range, to the whole human race, and even to beings of other races, whenever their existence has been clearly revealed. But it does not extend to all alike, however remote. The element of nearness or opportunity enters into the very essence of the command. In the law of gravitation we find one simple rule of insensible graduation. But here, in the great law of moral duty, there is a gradation of a far more complicated and various kind. All the relationships of human life, the laws and usages of social order, the conditions of time and place, and the sacred ordinances and bonds of religion, supply various elements, by which the love of our neighbour is more and more diversified; till it is enriched with colouring as various and beautiful as the hues and shades of light in the bow of heaven.

But it may be objected, finally, that even if moral truth in its own nature, from the definite character of its first axioms, and their capability of expansion, admits of scientific treatment, its immense complexity forbids the hope of any real success, when we pass beyond a few precepts, which sound like truisms, and attempt to unfold it in forms available for the conduct of life. What is most wanted in the present stage of Moral Philosophy, according to Professor Grote, is "not definiteness of system, but largeness of view." "Of course," he adds, "this renunciation of system, so far as it goes, lowers moral philosophy from its scientific rank, and alters it from the character of a single science to that of a group of sciences, whose relation to each other is not easy to determine." These remarks have doubtless an element of truth. They would be wholly true, if narrowness and definiteness were the same thing. He who looks, in clear sunlight, on a field shut in by high walls, has only a narrow view, but what he sees he sees clearly. He who tries to look on a wide, unbounded landscape, when a thick mist and fog is on every side, has a view, not exactly narrow, but indefinite and confused. The vision, practically, may be more confined and partial than before. The true system, in every field of science, is that which discerns clearly the relation of central truths, and traces them out towards the distant horizon; but also recognizes how dimly the more remote are seen, and how much lies beyond the horizon, and remains unknown.

In Moral Science, as in mathematical and physical study, it is thus of high importance to distinguish between principles and their application, the firmness and certainty of primary and fundamental laws, and the complex elements of man's earthly life, which diffract their pure light, and force on us the recollection of the weakness of our faculties, and our need of dependence on a higher wisdom than our own. The contrast in the words I have quoted needs thus to receive an important modification. In Pure Ethics our first and main want is clearness and definiteness of system. In Applied Ethics the great want is largeness and comprehensiveness of view. By the first we claim the privilege of moral eyesight. We escape from fog and mist and darkness, and gaze on a landscape, where we see, and know that we see, moral truths, sure and certain, which cannot deceive. By the second, we throw down the walls which would confine our view within too narrow limits, and see vistas of moral consequences, reaching out far and wide in all directions, which we see dimly and imperfectly at the best, and which lose themselves in the unknown.

It is common, in these days, to speak with exultation of the progress of science, and to boast of the many discoveries of physical students, by which we have far outstripped the attainments of every previous age. And no doubt this progress, in itself, deserves notice and admiration. But if Physics only advance, while Moral Science recedes and goes backward, our loss, we should find out ere long, would immensely exceed our gain. There can be scarcely, I conceive, a sadder spectacle, than an age of great intellectual activity, and of moral torpor and sloth; when men trace out busily and eagerly the laws of matter, and at the same time grow deaf to the claims of conscience, and blind to the supreme authority of the great law of love. It was the honour of Socrates to recall the thoughts of the Athenian youth from doubtful speculations in physics to questions affecting the moral duties of daily life. May not the like lesson still be needful, even in our Christian nation, and when Physics has risen out of doubtful and loose speculation into the real and progressive discovery of natural laws? Men may be eager to trace with prism and spectroscope the most subtle phenomena of light, and still be content to remain in moral darkness. They may search out, with line and dredge, the depths of ocean, and care little about the moral depths of sin and want in the human heart. They may infer, from subtle combinations of theory and experiment, the lessening distance of Sirius, and wholly neglect the revealed doctrine, echoed by natural conscience, of the sure approach of death, resurrection, and a life to come. They may be confident of the latest guess of theory on the fossil shells or moving glaciers of some distant, unknown age; and be satisfied to remain in doubt whether there is a God who made them, a law of right which they have broken, a moral pathway of hope and recovery open before them, and the solemn reality of a coming judgment. But to be positive about truths of the lowest class, and contentedly ignorant of the highest, to preach up science in the mere heel of the body corporate of truth, and canonize ignorance and endless doubt on those subjects which are the head and crown of the whole, can be no mark or sign of intellectual progress. It is rather the proof of growing anomaly and dangerous disease. He who formed the spirit of man within him has never made him capable of finding certainty in numbers and triangles, circles and polygons, and wholly incapable of knowing himself, and his own duties, hopes, and prospects, as a being made in the image of God, far nobler and higher than the fowls of the air or the beasts of the field. The maxim, γνῶθι σεαυτόν, which heathens recognized as a Divine gift, is far too precious to be thrown to the moles and bats by Christian men, under the vain fancy that sure truth is to be found in Physics alone. It may be hard and difficult to climb the hill-side, and we may sometimes, amidst the mists that surround us, be ready to abandon even the effort to rise. But to those who persevere the promise is sure, that clearer and clearer light will dawn around them. The scales of sense will pass from their eyes, and the mists will slowly disappear. That grand, comprehensive Moral Law, which it is the object of Ethics to discover and unfold, the true ideal of human thought and action, is no child of night and darkness. It finds its proper home

> In regions bright of calm and serene air,
> Above the smoke and stir of this dim spot,
> Which men call earth,—

those regions which are the native home of all truth and certainty, of light and love.

To this law, far more than those of physics, the noble words of Hooker will apply. "Her seat is the bosom of God, her voice the harmony of the world. All things in heaven and earth do her homage, the very least as feeling her care, and the greatest as not exempted from her power. Both angels and men, and creatures of what condition soever, though each in different sort and manner, yet all with uniform consent, admire her as the mother of their peace and joy."

LECTURE IV.

THE PRIMARY CONCEPTION OF MORAL SCIENCE.

THE definition of a Moral Agent must form the starting-point of all Ethical Science. Like the definitions of a point, a line, and a surface in Geometry, we must expect that it will touch on what is mysterious and hard to explain. But this mystery will form no adequate hindrance to a true and just conception, capable of preparing the way for a large superstructure of moral truth.

A Moral Agent is a person or individual, endued with the powers of action, feeling or sensation, and spontaneous choice; self-conscious, or capable of reflecting on his own actions, feelings, and volitions; and knowing good and evil, or apprehending a standard of good and right on one side, of wrong and evil on the other, with which his own thoughts, feelings, and actions, and those of others, may be seen to agree or disagree. It seems thus to imply the concurrence of three main elements. The first constitutes a sentient, the second a reflective or self-conscious agent, the third and last, a being conscious of a law of duty which he is bound to obey.

First of all, a Moral Agent is a person or individual being. The negative theory, which turns men and animals into mere bundles of sensations, tied together one knows not how, is fatal to the very existence of Moral Science. But it is fatal to Physical Science also. Until we rise from separate phenomena to the conception of things and persons, by which these are caused, and to which they belong, we stick fast in deep mire where there is no standing, no foothold for the human reason. Let us adopt, for instance, the definition of life which a recent speculator has proposed, that "it is a definite combination of heterogeneous changes, both simultaneous and successive, in correspondence with external coexistences and sequences." Here we have an attempted portraiture of life, where there is no recognition of any living thing or person whatever. We have a series of changes, and nothing more. It is the play of *Hamlet*, where no *Hamlet* is to be found. But our first conception of life is always that there is some one being which lives. No series of changes, whether simultaneous or successive, whether homogeneous or heterogeneous, can possibly give the conception of life, unless they first suggest the perception of some living thing or person, insect, bird, beast, or man, to which they belong.

Again, every living thing suggests and includes the ideas of action, sensation, and spontaneous choice. Whatever is alive acts, and it also feels. But we also ascribe to it some kind of choice or spontaneity. The elements which enter into this choice may vary widely. But even in the lowest forms of life we conceive action to depend in some way on the feeling or sensation of the living thing, and not to be decided directly and simply by external impulses alone.

A living thing, however, may act and feel, and in a certain sense may choose how to act, and still be incapable of reflecting and reasoning upon its own actions. With the power of reflection or self-consciousness we pass from the animal to the human or rational stage. The very name, Conscience, from *conscire*, implies a duality, in which the thinking being is its own object of thought, lives, in some sort, side by side with itself, and not only has such and such powers, but is conscious of them, and can reflect upon them. This power of self-knowledge, in which the soul can see itself, compare its own feelings, judge of its own actions and thoughts, and through its own self-consciousness come to recognize the existence of conscience in others, seems to distinguish man from the lower animals, and is essential to the conception of a Moral Agent.

The word, I, in contrast to all splendid and pretentious polysyllables, has thus been truly placed by Prof. Maurice as a kind of sentinel at the gate of Moral Science. It is a key, he justly remarks; "to that mystery in words, which makes them interpreters of the life of individuals, of nations, of ages; the discoverers of that which we have in common, the witnesses of that in each man which he cannot impart, which his fellows may guess at, but which they will never know."

But a Moral Agent is, further, one who knows good and evil; who can not only reflect on his own actions and feelings, as his own, but compare them with an apprehended ideal of the good and right, and discern, in his own case, and in that of other moral agents, their agreement or their disagreement. Thus self-consciousness rises into a conscience of right and wrong, of good and evil. The comparison will be latent and potential when there has been no inward experience, or outward observation, of evil and wrong. It becomes explicit and clear, whenever evil has been witnessed or experienced, as in the sacred narrative of the first human trangression,

Knowledge of good, dear bought by knowing ill.

Every sentient, self-conscious being, capable of recognizing an ideal standard of right and wrong, of good and evil, and of being guided in thought, feeling, and action, by motives drawn from the perception of such a standard, is a Moral Agent. He is not only an I, but an I who says to himself, This I ought, and this I ought not to be, to feel, and to do.

The sense of Moral obligation, in every such agent, depends on the union or concurrence of two elements. The first is the power of voluntary choice itself. The second is the capacity of perceiving, in various forms of feeling and action, the moral contrast of good and evil, of duty and transgression, of right and wrong, and of being influenced by motives of this higher kind.

The vexed question of the freedom of the human will enters thus into the very foundations of Moral Science. But it is important to observe the practical limit of the connection between them. We do not need fully to resolve a deep problem, which loses itself in mystery, how far the will is determined by motives, or itself fixes the weight and decides the relative order of the motives themselves. All we need to know is that there is an extreme on either side, which is fatal to moral obligation. Within these limits, our defective knowledge of the nature and actings of the will does not hinder the sense of duty from remaining firm and strong.

Once let us suppose that choice is an illusion, that human acts are decided, like the motion of a planet in its orbit, simply by the concurrence and resulting effect of many external impulses and physical sensations, and then the sense of responsibility, or of a moral character in our thoughts and actions, disappears. Men would then have no more claim to be moral agents than the falling apple or the flashing meteor, than stocks or stones, or the waves of the sea.

On the other hand, suppose the will to act without any motive or reason whatever, by a blind caprice, an impulse without an aim, and responsibility would cease on the other side. Such an agent would be like the fictitious atoms of Epicurus, that turn aside a little from the right line, without any reason why they deviate one way more than another. But he must cease thereby to be a reasonable being. He becomes an embodied chance, a capricious and senseless atom. For surely to act even on a mistaken motive, and from an insufficient reason, is at least one step higher than to act with no motive or reason whatever.

Between these two extremes, of a volition purely capricious on one side, and of mechanical, physical compulsion on the other, there is a wide interval where moral responsibility has its natural home. Man is not so free as to be able to dispense with motives altogether. Yet he is bound by no such chain of fatal necessity as that motives are mere mechanical forces, and can decide his course irrespective of his own inward state, and the character of his whole moral being. How far motives, by their habitual prevalence, form the character, and how far pre-existing character for good or evil varies and affects the force of the motives, are deep and hard problems of the human heart. In these we may be able to see but a little way. But the human conscience, whenever it is not bent on vain self-excuse, and drugged by some fatal opiate, decides at once that these limits do obtain in the actual constitution of human nature. Man is still free enough to be responsible, however grievously, in many cases, he may be "holden with the cords of his sins."

And still he is not free enough to despise with safety the double warnings of experience and of revelation on the power of vicious habits, and the bondage to which they swiftly lead. Those words of the prophet clearly express a solemn and weighty truth,—" Can the Ethiopian change his skin, or the leopard his spots? Then may ye also do good, who are accustomed to do evil."

The power of reasonable and spontaneous choice within these two limits is the ground and basis of moral obligation. The two ideas, like the convex and concave in one and the same curve, are correlative and inseparable. The truth has its inward aspect, towards the agent himself, and his own self-conscious power; and its outward aspect, towards the law of right above him, and the whole moral universe. Every where and in all things Law must reign, and chance is only a shadow. If Man, then, by reason of his higher powers, escapes from the control of a physical necessity, and from laws he cannot help obeying, he must come thereby within the range of a higher law of right and duty, which speaks to him in nobler accents with the voice of a sovereign, and which he knows and feels that he ought to obey.

The close connection of these two ideas, power of reasonable choice, and moral obligation, is doubly confirmed by the cases they include, and by those which lie just beyond their limit. Wherever that power is recognized, however wide the other differences, we own the clear existence of moral duty. Thus the existence of angels is revealed to us in Scripture. Of their special powers, their gifts and modes of intelligence, very little, and almost nothing is known. Still they are clearly set before us as intelligent beings, endued with the powers of reason and choice. Therefore, while ignorant of nearly all beside, we recognize at once that they are moral agents, bound by the great laws of moral duty.

Again, philosophers have pleased themselves with inventing conjectural races for other planets, endowed with reason, but in other respects supposed to differ widely from mankind. But these inhabitants of Venus, these Jovians or Saturnians, are no sooner invested in our thoughts with will, choice, and reason, than we feel that the laws of duty must also apply to them. The same is true of every Eastern tale of magical change of form or supposed transmigration. The magic which changes the shape alone leaves the sense of moral being unaltered. But the moment we suppose reasonable choice to be gone, and blind instinct alone to rule in its stead, the whole moral nature has disappeared.

The same connection reveals itself, no less clearly, when we dwell on those cases which seem to border on moral obligation, without fully attaining it. Three such cases may be specified—children, lunatics, and those domestic or nobler animals which attain some striking resemblance to the virtues and vices of mankind.

The question is often perplexing, at what stage of its life a child first begins to be responsible. So long as the infant merely continues and multiplies its experiments on the outward world, we do not ascribe to it a moral character. As soon as it gives signs of self-reflection, we feel that in that childish bosom a nobler life has begun to be revealed. And whenever this reflection seems to involve a power of comparing its acts and feelings with a law of duty, it becomes plain that the exercise of a moral nature has begun. We place it then at once in the higher category of moral and responsible agents. Those infants only escape from the claims of this law, which, in the words of Scripture, have no "knowledge of good and evil," no sense of a standard of right above and beyond themselves.

The case of lunatics seems, at first sight, more perplexing. There is here not only frequent embarrassment, but a strong tendency to mutual divergence between doctors on one side and lawyers and divines on the other. The former are prone rather to extend, the latter to contract, the range of exemption from moral responsibility through the partial failure of reason. The perplexity arises, I think, in no small measure, from confounding pure Ethics with that mixed, composite morality, which is involved in human jurisprudence. The lunatic, in general, does not cease to be capable of strong convictions about moral right and wrong. But, owing to some mental illusion, the connection which may be assumed in other men between certain actions and the answering motives is distorted and deranged. His inward feelings do not lead to their natural and healthy results. His outward acts, when tried by customary rules, do not indicate, with any clearness and accuracy, his inward motives. Thus he does not cease, for a moment, to be inwardly responsible in the sight of God. But the common rules by which human law infers crime and guilt in the heart from the outward actions fail to apply, and if they are enforced, would lead to serious injustice by their application. The method of solution which replaces all forms of punishment by simple detention seems to be rather an idle and clumsy way of cutting the knot, which it would need much care and pains to untie. In avoiding wrong to the individual, it runs the risk of serious danger to society, and of a general relaxation of human law; for all vice and crime is highly unreasonable, and therefore, in one sense, a presumption of partial madness. Lunatics have been known to attempt or threaten murder on the ground of their own expected impunity through mental disease. Surely the true policy, in cases where criminal acts, by the verdict of a jury, are connected with strong suspicion or proof of partial derangement, would be to refer the decision on the partial or entire remission of the legal punishment to a mixed tribunal of lawyers and eminent physicians, selected and permanently appointed as a court of special jurisprudence.

Such cases, however complex and difficult in practice, only bring out the principle itself into clearer relief. So far as involuntary illusion divorces the outward act from the motive it would otherwise infer, the action ceases to wear the moral character which the law, in other cases, justly assigns to it. But, so far as the mind can choose and reason, though partly under false impressions, it continues to be morally responsible. The difference is that more exact care and closer inquiry is needful, to decide on the moral elements of the inward feelings and intentions, as deduced from the outward acts. An oar does not cease to be an oar, because, when dipped in water, it seems to be broken. Allowance needs only to be made for the diffracting power of the medium in which part of it is seen. The lunatic continues responsible for his own motives, feelings, desires and intentions. He is not responsible for his outward acts, so far as they are turned aside, by involuntary illusion, from the course to which these inward desires and feelings would lead in a healthier mind.

Another subject, which tends to obscure the doctrine of moral responsibility, is the close approach, at least, to the characteristics of moral vice or virtue, which some animals appear to attain. How natural it is to apply moral epithets of praise or blame, to the dog faithful to his master even to death, to the vicious, unmanageable horse, the generous war-steed, the cunning serpent, and those various domestic favourites, which seem to come within the shadow of human influence, and assume, in some measure, the features of a higher nature than their own!

On the strength of this evident fact an attempt has been made to found the theory that all moral feelings and emotions are only animal instincts, slightly transformed. The view falls in with the doctrine of successive evolution of species, so popular at this hour with a large number of physical students, and espoused by them almost with the zeal of a new religion. The moral sense is said to follow "firstly, from the enduring and always present nature of the social instincts, in which respect man agrees with the lower animals; and secondly, from his mental faculties being highly active, and his impressions of past events extremely vivid, in which respects he differs from the lower animals. Owing to this condition of mind, man cannot avoid looking backward, and comparing the impressions of past events and actions. He also continually looks forward. Hence, after some temporary desire or passion has mastered his social instincts, he will reflect and compare the weakened impression of such past impulses with the ever present social instinct; and feel that sense of dissatisfaction which all unsatisfied instincts leave behind them. Consequently, he resolves to act differently for the future, and this is Conscience. A pointer dog, if able to reflect on his past conduct, would say to himself, as indeed we say of him, I ought to have pointed to that hare, and not have yielded to the passing temptation of hunting it."

A pointer dog, "if able to reflect on his past conduct," would be a highly intelligent, almost human being. And, if able further to say to itself, "I ought to have resisted that passing temptation, which turned me aside from the law of my duty, and to have pointed at that hare, instead of hunting it," such a dog would plainly be a conscientious being, a moral agent. But the "if" and the "I ought" in this hypothesis are instances of what has been wittily called "the Saltatory Principle," which some have called in to supplement the doctrine of slow development. It leaps over an immense gulf at a single bound. Compared with the Man described just before, who merely seeks to gratify a suppressed and reviving instinct for social pleasures, the dog would rank far higher in the moral scale. The attempt to bridge over the interval between man and the brute is ingeniously made by a double process of change. Man is degraded into a slave of the stronger instinct, and his blind obedience to it is dignified with the name of conscience. The animal is promoted into a self-reflecting, conscientious moral agent; who considers his own ways, and says to himself, like a moralist or divine, "This is a temptation to which I ought not to yield. This, and not that, is what I ought to do."

The only force in an argument doubly fallacious consists in the undoubted fact that the conduct of animals, in some cases, does spontaneously suggest the use of moral epithets. It is natural, almost unavoidable, to apply to them words of praise or blame. How can this be explained, if the power of a reasonable choice and moral responsibility are inseparably linked together?

Now in what cases does the conduct of animals rise into what we feel, at least, a very close resemblance of moral virtue? In every case, I think, it is when some lower instinct, either of indolence or animal pleasure, yields to the higher motive, obedience to a superior will. The faithful dog finds its chief delight in pleasing its master. It listens to his voice. It watches his eye. It follows promptly, and with evident delight, the slightest indication of his will. Endowed with senses and powers of delicate scent or swift motion, superior to his master's, he still takes pleasure in lending them to the service of one in whom he feels instinctively the presence of a nature higher than his own. It becomes to him a second nature, strengthened by a thousand acts, that his master should command and he should obey.

All moral action consists in obedience to a law of duty, which appeals to the highest part of man's nature, even at the sacrifice of lower instincts and momentary pleasures. When united to religious faith, it consists in the recognition of a higher, uncreated Will, supremely good, and in obeying that Will, so far as its lessons are known and understood, at whatever price of self-denial, pain, or suffering.

It is plain, then, that all these instances of devoted attachment, on the part of animals, to which epithets of moral praise are commonly applied, bear the very closest analogy to virtuous actions in moral, intelligent beings. The dog, which looks up fondly to its master, and obeys every sign of his master's wish, follows within the limits of its own power the very same law which constitutes piety and moral goodness in human beings. He practically apprehends and follows his own highest good. His actions are not only naturally good, they border closely on moral goodness. They bear to it the closest analogy and the nearest resemblance. Let us only, by a tacit illusion, add to them the idea of something higher than instinct, of a free and reasonable choice, and they become instances of moral goodness in one of its purest and noblest forms.

How, then, are such acts of animal nature to be distinguished from actions properly of a moral kind? By the want of power in the animal to rise beyond the visible standard, the will of its master, and to compare its actions with a standard of absolute right still higher. A criminal, we may suppose, the owner of a faithful dog, to escape from the police, stirs it up to attack them, and it is killed in defending its master from the hands of justice. We admire and praise its fidelity, to which its life is sacrificed. We impute the blame and condemnation to the vicious owner alone. To obey the human will is the dog's highest law, and it seems to us incapable of climbing higher. When an instinct so noble is perverted and abused through the sin and folly of man, we have simply a proof that man, when he deserts the true standard of his being, is not only made subject to vanity in his own person, but also drags down with him the lower creation in his grievous fall.

But there seems to be a further truth involved in facts of this kind, and which is needed to complete the explanation. There is a singular class of facts in the science of harmony, called the doctrine of forced vibrations. Notes or strings of a particular pitch or length, when their sound is powerful, tend to hinder and quench those vibrations of other strings which are in discord with them, and awaken others which are either in unison, or bear to them the relation of some fundamental chord.

There seems to be a similar law in ceaseless operation in the world of thought and intelligence, when lower natures are brought under the daily and hourly influence of some being of a higher kind. They are attracted by a kind of magnetic influence. Their own instincts are more or less suspended, and replaced by others of a higher sort, the echoes and reflections of that higher nature which dominates over them. They are thus magnetized into the semblance of a higher life than is properly their own. And thus domesticated animals, in proportion as the society of man has its full influence over them, seem to rise into a kind of border land between mere instinct and moral intelligence. They are raised by a foreign power above themselves. A process of elevation is at work, which tends to conceal the interval between instinct and reason, unmoral and moral life, because the lower instinct is guided, controlled, and in a manner subdued and replaced, by a power derived from man's higher and nobler being. The strange marvel, recorded once alone in Scripture, and even there a stumblingblock and jest to the careless and profane, would seem thus like one extreme instance of a general law in ceaseless operation. It may be taken for an outward visible sign of a sacramental privilege of moral elevation, gained by lower creatures through the daily consecration of their best powers to the service of man. When the prophet was perverse and faithless, the ass which had served him faithfully so long, rose for a moment to a share in its master's abused and forfeited privilege. It saw what the man whose eyes had been opened could not see. The dumb ass, speaking with human voice, forbad the madness of the prophet.

The faithful animal, which cleaves to its master with a fond affection, and sacrifices its own instincts in obeying his will, attains thereby the highest type of its own being, and a close resemblance to moral goodness in beings of a higher type than its own. And when this analogy is confirmed and deepened by such daily and hourly intercourse that the higher, intelligent will seems to mould and almost animate the lower, the resemblance is closer still, and seems almost like a direct participation. But these border cases only confirm the truth of the principle, that a power of reason and free choice, and this alone, is joined with the sense of moral obligation. They prove that instinct in its highest and noblest forms is, in the words of Coleridge, like "the shadow of approaching humanity, the sun rising from behind, in the kindling dawn and morning of creation."

But the same principle, the primary and fundamental law or conception of Moral Science, precludes at once, and wholly sets aside three counterfeit theories of conscience, the physiological, the instinctive, and the tyrannical. Let us examine these briefly in succession.

The Physiological Hypothesis in Morals consists mainly in the union of two elements. The first is the process of medical research, by which the consequences of all mental emotions are traced in their effects on muscular and nervous tissues, and on the state of the bodily frame. The second is a vague, ill-defined theory of the perpetual equality of dynamical force, under forms ever varying. All moral emotions, it is thus inferred, are only the results of some strange concentration and transmutation of solar force in the nervous structure of the human frame.

Now it is plain at once that since we ascribe moral responsibility to angels and other possible unknown races, of a physiology wholly different from our own, or bodiless spirits, and deny it to animals, however nearly allied to us in structure, so long as we exclude from our conception the gift of reason, the hypothesis must be an entire delusion. It excludes from the definition of a moral agent what is alone essential, and includes what is accidental and superfluous. Things may actually concur, and that in the whole range of our limited experience, and still we may feel clearly that there is no essential connection between them. Certain changes in the nervous tissues, let us grant for a moment, accompany the moral emotions of love or hatred, sympathy or envy, hope or fear. How can this prove that the physical change is the cause of the emotion, and not its effect? How can it prove that the connection is not wholly arbitrary, and capable of being entirely different for other races, or for men themselves in the life to come? Thus, for instance, we cannot think of a circle, or reason upon it, without some sensible image. There may be many to whom it always suggests the thought, by association, either of a plate in a dinner service, or of the sun in the sky. But if some skilful manufacturer should try to persuade us, that the properties of the circle depend on a knowledge of the earths that are used in making porcelain, or the spectroscopic observer, that they rest on a chemical analysis of the red protuberances of the sun, then, in spite of all their manufacturing skill or scientific attainments, we should justly laugh to scorn so prodigious a folly. And it is and must be a folly no less extreme, when the attempt is made to resolve into some complex medical fact about the waste and change of nervous tissues, those grand fountain truths of moral good and evil, which encompass, guide and sustain the whole universe—those laws of eternal goodness and righteousness, which are the deep foundations of the throne of God.

But the mere instinctive theory of morals is equally set aside by the same test. A man has lost favour with his fellows by some selfish or cruel act, the result of self-will or animal impulse. He is thus disliked, and his social instinct is troubled and dissatisfied. To get rid of this pain, "he resolves to act differently for the future, and this is conscience!"

Conscience, according to this new and strange definition, is nothing more than an instinctive effort to escape from the secret pain of an unsatisfied desire. It would seem to follow that every one is conscientious, in proportion to his eager attempt to procure the unbridled and ceaseless indulgence of every lust. The great author of evil, according to Milton, must have had his conscience in vigorous exercise, when he set out on his expedition to deceive and destroy.

> "Gabriel, thou hadst in heaven the esteem of wise,
> And such I held thee; but this question asked
> Puts me in doubt. Lives there who loves his pain?
> Who would not, finding way, break loose from hell,
> Though thither doomed? Thou wouldst thyself, no doubt,
> And boldly venture to whatever place
> Farthest from pain, where thou might'st hope to change
> Torment with ease, and soonest recompense
> Dole with delight, which in this place I sought."

And if a strong social instinct is needful to complete the definition of conscience, this too is satisfied in the description of his eager longing for closest "league and amity" with Adam and Eve and their offspring, which the arch-fiend has professed just before in his famous soliloquy. Our great poet hardly could have suspected that a philosophy would arise, which would find all the needful elements for the genesis and birth of conscience in his own masterly portraiture of the Prince of darkness, and the vehement workings of self-tormenting and consummate evil. Avoidance and flight from evil, and craving for companionship at whatever cost, are seen there united in full perfection.

Near akin to this theory of conscience, as generated by the effort to escape from an unsatisfied instinct, is the tyrannical theory, which sees in it submission to the will of the stronger, and a mere result from the habitual dread of human punishment. On this view it is an artificial product and creation of the penalties which lawgivers have imposed. Thus a strong association is produced between certain lines of conduct, and the dangerous and painful consequences to which, under the actual constitution of society, they naturally lead.

Now it is doubtless true that by means of human laws and their penalties we rise most easily to a vivid perception of moral right and wrong. We are told, on the highest authority, that "by law is the knowledge of sin." It seems, however, no less plain that human laws do not create the conscience of right and wrong, but only awaken it. Men come very soon to distinguish between actual laws, and a higher standard of right, by which laws themselves need to be measured. It is felt that legislation, in some cases, may be unwise and mischievous, and that punishments, though in agreement with the letter of human laws, may be unjust. How can water rise above its source? How can the conscience of man, if it be merely a product of the fear of punishment, ever pass sentence on those very laws by which punishment is enjoined? The authority which such laws carry with them depends plainly on two causes. Either what they enjoin is seen to be right in itself, and what they forbid to be wrong; or else the relation between those who command, and others to whom the command is given, is felt to involve the duty of obedience, whenever the command is not in its own nature morally evil. In both cases alike we are compelled to recognize a source of obligation higher than the mere will of the lawgiver. In one case it reminds us that such things were duties, and right to be done, even before the human command was given. In the other, the higher law enjoins obedience, within certain limits, to the known will and command of a lawful superior. But without some faculty of moral discernment, some standard of right deeper and higher than human laws, the laws themselves must be wholly powerless. Nothing would be left of them, but the power of the stronger to inflict suffering on the weaker, whenever there is not abject submission to their will; and the exposure of the weak to be crushed by the strong, till their turn may come to cast off a hateful yoke, and become oppressors in their turn.

Once let these theories of morals be condensed into their simplest form, and stripped of those accessories of ingenious thought, which disguise their true features, and they will be spontaneously rejected as wholly incapable of explaining the real experience and deepest convictions of the human heart. When a man has committed a frightful murder, his whole nervous system undergoes a strange and painful disturbance, and this is conscience! When he has forfeited the esteem of his fellows by selfishness and vice, his social instincts are unsatisfied. He resolves to gratify them in future, and this is conscience! The slave crouches with fear before a human slave-driver, and shrinks from the lash he sees uplifted, and ready to fall, and this again is conscience! Yes, they are conscience in the same sense that the monkey, the chimpanzee or the gorilla is a reasonable man. There are some strong features of bodily or anatomical resemblance, but the human heart and mind are wholly wanting. Such a conscience is no true conscience, but a poor and wretched substitute where the reality is wholly absent. A murderer may experience strange horrors. The "horrid, horrid dream" of the scene of violence may beset him in his waking hours. He may have the strongest desire, and the firmest resolution, to disguise his crime, that he may still enjoy the pleasures of social intercourse. He may be full of secret terror, lest his course should be detected, and the sentence of the law alight upon him. And all the time conscience may be asleep. He may remain a hardened and cruel ruffian, ready, if only he can reckon on impunity, to repeat the worst of crimes.

True Conscience is a higher and nobler thing. Its ground and root is the conscious possession of a reasonable will, an inward power of choice, dependent on no mere external influence, but on the mind's own conviction of what is fit and right to be done. Its upward growth is towards the light, sometimes dimly seen, as in mist, sometimes as in brighter daylight, of some noble and lofty ideal of human conduct, some bow of heaven, in which the pure light of a perfect love diffracts and varies itself into a thousand forms of loveliness and beauty. It lifts the eye of the soul towards those eternal hills, whence its help for moral conflict and victory must come. The physical elements of man's bodily frame do not and cannot enter into its definition. It belongs to the reasonable mind and will, whether unembodied or disembodied, to angels and to men, to men living as we now live, or unclothed by the stroke of death, or clothed upon with nobler bodies in the resurrection. It deals with voluntary actions in their threefold character, the source from which they flow, the limits within which they move, and the issues to which they tend. But moral tendencies cannot even exist, when moral characters and distinctions, in the will which is the fount of action, are wholly denied. The tree must first be good, that its fruit may be good also. There must be some character of moral goodness in the motives, tempers, and feelings of the heart themselves, before they can have any tendency, in their actings, to produce good even in distant ages, and throughout the whole extent of the moral universe.

The law of gravitation, in the natural world, is wonderful in the consequences it involves, and the results to which it leads. We live, it tells us, surrounded by mysteries without number. All creation is bound by it into one great system of being. Every leaf that withers and falls in the forest, affects, though insensibly, the motion of every planet in that great system. The dew-drop, alighting on the grass, has its speed increased by every gem buried in the Indian waves, and every wave that dashes on the beach of a thousand distant isles. Links of secret influence emanate from our own bodies, from every bird and insect in the skies of summer, and range onward, till they lose themselves in the depths of space. There is nothing solitary in the wide creation. There is no single mote or atom, in earth or air or sea, but is linked fast, by relations science has in part revealed, and which in part are still hidden, to the whole system of created and material things.

Now conscience discloses to us a law of union in the moral world, as wide and far reaching as the law of universal attraction, but of a nobler and higher kind. It is a law, not of physical compulsion, but of moral duty. It pronounces, not what will be done, but what every moral agent ought to be and to do. Revealed to us more dimly by natural conscience, it stands out in still clearer light in the word of God. But clearly or dimly taught, the message is the same. The sacred voice may be repeated, but the vision is one. That law is the great and simple law of love. It includes love to God, and love to men and all the creatures of God. It ranges through every rank and order of sentient beings, as the objects it includes within its care, though moral, intelligent, reasonable creatures are the only subjects to which it belongs. It stoops to protect the insect from the lawless cruelty of the child, and rises to claim the allegiance of angels and archangels before the throne in heaven. It seems at first, like the pure sunlight, to be so simple that nothing can be learned from it, and thus it may be in danger of being mistaken, by the rash and careless, for an empty truism. But when it stoops down to earthly things, and passes through the clouded atmosphere of this lower world, it discloses an infinite variety of rights, duties, and obligations, which belong to men in their relations to each other, to the unseen world beyond the grave, and to the great Author of their being. And when it has ranged through the wide universe, and stooped down to awaken high and pure desires in hearts that have been deluded and darkened by vice and folly, recalling them to the upward pathway of light once more, it returns to present its gathered wealth of all holy desires, all good counsels, and all just works, as a freewill offering at the footstool of that glorious Fountain of goodness, perfect in all moral beauty and holiness, from whom they proceed, and to whom they tend as their proper home. For "God is love, and he that dwelleth in love dwelleth in God, and God in him. He that loveth not knoweth not God, for God is love!"

LECTURE V.

THE DIVISIONS OF MORAL SCIENCE.

ALL science, as Lord Bacon truly remarks, has three main subjects, Nature, Man, and God. There are thus three main divisions of human knowledge, Natural Philosophy, Humanity and Theology. Natural Philosophy, again, is simple, because in material things, and mere vegetable and animal life, there is no standard of right and duty, distinct from their actual state. Theology is also simple, because in the great First Cause, what is and must be, what is and what ought to be, through the perfection of the Divine Nature, are one and the same. The true conception of a God includes the highest conceivable perfection. But our knowledge of Man and Moral Agents is twofold. Side by side with the lesson of their actual state, derived from experience, we are compelled to recognize an ideal standard, a law of duty and perfection, which they may or may not attain. And hence there is a twofold Science of Humanity, the Actual and the Ideal. Ethics is the science of Ideal Humanity. It is thus coextensive with the wide range of experimental or practical human science, and presides over it, as it presides over Physics, being the lawful superior and mistress of both. And it borders on Theology, of which, rightly understood, it is neither the rival, the child, nor the slave, but the spouse and ally, subordinate in holy and loving union. For in reality it is only through some awakening of natural conscience that men can rise to the discernment of spiritual truth.

Ethics, again, have science and certain truth, not mere guesswork and probability, for their aim. Such a view is confirmed, not only by the judgment of Locke and my two most eminent predecessors, but by the nature of the subject, and the analogies of geometrical science and physical research. It must be possible, with humble and patient thought, to rise here above the mists and fogs of endless doubt into a higher region of clear sunlight and eternal truth. The definitions are simple, on which the whole superstructure depends. The whole temple of moral truth may be said to rest on these two pillars— a power of reason, will, and spontaneous choice in each moral agent, and a perceived standard of right and wrong, binding on such a free moral agent, as his true ideal of thought, feeling, and action, by its own inherent and supreme authority. That standard, reduced to its simplest form, and summed up in one word, is the great law of universal love, or the first and second great commandments of the Law and the Gospel, bound together in essential, indissoluble, and perfect union.

Such a view of the nature of Moral Science may seem perhaps, at the first glance, to reduce it within the narrow limit of a few barren and almost self-evident truisms. But such an inference would be a fatal misconception. The star we see at midnight may appear, to the careless observer, to be only an insignificant speck of light. But the seeming minuteness results from its distance alone. The sun is a million times larger than the earth on which we live. Sirius, from modern photometry, has been inferred to be sixty or a hundred times intrinsically brighter than the sun. It may thus, for aught we can tell, be the centre of a material system, vaster and larger than our own. So also the simplest truths of science, like the properties of the triangle or the circle, or the law of gravitation, are the most fertile, when we strive to trace out the consequences involved in them, and the results to which they lead. And as soon as we endeavour to map out the fields of Moral Science, and to form the outlines of its geography, it will become plain how immense is the range of thought which has here to be traversed. A rich and large variety of moral precepts, lessons and doctrines of high importance, will then be seen to unfold themselves from its fundamental truths.

Moral Science admits of being classified in three different ways. The first of these depends on the various objects or persons whom it includes; the second on the faculties and powers of the mind, which have to be brought under the control of moral obligations; and the third on those moral capacities, which are shown, by the joint evidence of experience and revelation, to belong to the whole universe of moral agents.

I. First of all, Moral Science may be viewed with reference to the objects or persons whom it includes within its range. It is thus parted at once into three main divisions, Personal, Social, and Divine. The third and last of these answers to the first great commandment of the law. The second, which is like the first, includes the two others. Theology, based on a supernatural message from heaven, descends naturally from the love of God, as the primal duty, to the love of man. But when we view Morals as a human science, the natural order is a climax, in which we begin with ourselves, and then travel outward and mount upward. Self-love, the love of our neighbour, or of mankind at large, and the love of God, are thus the three main divisions or kingdoms of Moral Science, as defined by the objects of duty which it includes, and the persons to whom it extends.

Self-love is the first of these three main laws of moral duty. Its just expansion and development open a very wide field of thought. If we are not to love ourselves at all, the command, Thou shalt love thy neighbour as thyself, is robbed of all sense and meaning. It would then constitute a virtual prohibition of every kind of social affection. Personal and social morality are thus inseparably linked by the Divine law. They rest on the same foundation, and must stand and fall together.

This branch of Ethics has been exposed to a double perversion from opposite sides. One class of writers would exalt it beyond its due limits, so as to constitute the whole of Moral Science. Others fall into an opposite extreme, and would exclude it altogether.

All the selfish theories of Ethics come under the first description, or the school of Epicurus, Hobbes, Mandeville, and the lower class of utilitarians, who have not changed their wild olive by grafting it on the more fruitful stock of Christian benevolence. To the opposite pole belong some mystics of the middle ages, the school of Kant, and some few intuitive moralists of modern times. According to the first class of writers, all morality is only a wise expediency, a prudent, foreseeing regard to our own happiness and enjoyment, and a wise calculation of the means of securing it in the fullest measure. According to the other, virtue must be pursued simply for its own sake. Any respect to our own benefit in upright or virtuous conduct degrades it from its proper character of obedience to an absolute law of right, and thus poisons duty at its fountain-head.

The pathway of true wisdom lies between these two opposite extremes. Those benevolent affections, which alone satisfy the wide and large claim of the great law of duty, are more than mere self-love, covered with a thin disguise. The mind, when in a right and healthy state, goes forth directly towards its object. It seeks the good of the loved one for his or her own sake, and does not pause first to make a secret calculation that their benefit will redound in probable or certain gain to ourselves. The benevolent affections, as Dugald Stewart has well observed, "prompt us to particular objects without any reference to our own enjoyment." There is no doubt that actions of outward beneficence may often proceed in their secret source from self-love alone. But then in this case they fail to satisfy the first conditions of social morality.

To the same effect Cicero remarks on the doctrine of Epicurus (*De Fin.* II. 24), "Where can there be a place for friendship, or who can be a friend to any one, whom he does not love 'ipsum propter ipsum,' himself for his own sake? What is it to love, but to wish any one to be enriched with the greatest benefits, even though there should be no return from those benefits to him who desires them? But it benefits me, you may say, to be of that disposition. Nay, perhaps, to *seem* to have it. For you cannot *be* such, unless you *are* such. And how can you be such, unless that love itself has possession of you? And this comes to pass, not by introducing the conception of its usefulness, but it is born of itself, and springs up of its own accord. But you say, I follow utility. Thy friendship then will last, so long as some gain shall follow it, and if utility makes a friendship, the same will unmake and destroy it."

The reasoning of Kant, on the other hand, would exclude from morality all pursuit of personal happiness. Our duty, he affirms, is to seek our own perfection, and our neighbour's happiness, but neither our own happiness nor our neighbour's perfection. This is a strange paradox in a writer so subtle and profound. It replaces the Divine command by another, which seems its exact converse— "Thou shalt love thy neighbour, but not as thyself." What every one inevitably wills, he says, cannot fall under the notion of duty, so that it is an intellectual contradiction to say that a man is obliged to advance his own happiness with all his might. But surely an instinctive craving for momentary pleasure differs very widely from a deliberate, rational pursuit of our own highest welfare. The first may sometimes be synonymous with open vice, and oftener still with careless and guilty folly. But the second, though it may have a further and wider range, must evidently include the aim at moral perfection. And thus it must ever form one essential element of true morality, as approved by the enlightened conscience, and directly enjoined in the Word of God.

Social Morality, the second main division of Ethics, has its own difficult problems. What is it to love our neighbour as ourselves? Does it mean that we ought to love him with exactly the same kind of love? In this case, if our self-love be excessive, diseased, and impure, we should have to extend the range of this disease and corruption, so as to infect the whole range of our social action. If we love ourselves inordinately and unreasonably, without candour or equity, the evil would only be aggravated by exercising an affection of the same kind, even if that were possible, to all around us. Or does the precept mean that we are to love them exactly in the same degree? This is hardly possible in itself. And it is an idea, which seems further excluded by the word, "neighbour." For this naturally implies that nearness enters into, and forms one element in determining, the moral obligation. The true meaning of the command seems rather to be, that the thought of self or not-self is excluded in estimating the relative claims of duty, and that the obligation rests on nearness, neighbourhood, and opportunity alone. We are nearer to ourselves than to others, and have thus fuller opportunities to seek our own good, and to act for it, than for theirs. But beyond this contrast, which cannot be set aside, and is of great practical moment, the rule prescribes that the question, whether the good be theirs or ours, shall not disturb the impartial actings of a wise benevolence, which will thus be found in perfect harmony with a temperate and healthy self-love.

The third and highest branch of Ethics, Divine Morality, or Moral Theology, gives rise to deep questions of another kind. If love consists in a genuine desire for the happiness and well-being of others, or of ourselves, how can it apply to One, of whom not every Christian only, but every intelligent Theist, must believe that He is purely and perfectly blessed? But the answer to this difficulty is not hard to find. Love may assume three different forms or aspects, according to the diverse state and character of the object beloved. It may rest on those who are in want, pain, suffering, and moral degradation. In their case it assumes the form of pity and compassion, of desire for their moral recovery, or their relief from suffering. Again, it may have respect unto those who are upright, sinless and happy, but whose happiness is limited, and their moral excellency still capable of a large increase. In their case love must assume the form of simple benevolence, of desire for the increase of their happiness, and their growth and progress in all things that are good and worthy of praise. But, lastly, love may have for its object the glorious Creator and Preserver of the universe, the Infinitely and Supremely Good, who is blessed for ever. And here it must pass into a higher and nobler form. It becomes a delighted and adoring complacency in the contemplation of His infinite goodness. Divine Love is the exact reverse of selfish murmuring, or of sinful envy. It implies a deep gladness of heart, that He who is perfectly good and holy should be removed far above the reach of pain and evil, and that He should be blessed with perfect and inconceivable blessedness for evermore.

II. Moral Science may be distinguished and arranged, in the second place, by a reference to those faculties or powers of the mind and will, which are seen to coexist in every moral agent. The threefold character of the science meets us here in another form. Goodness, wisdom and power, all infinite, meet in our conception of the Divine Being. Being, knowledge or intelligence, and will, or the *esse, posse* and *velle* of the schoolmen, are equally found in the constitution of every moral agent. So that Moral Science must have a threefold character, when viewed in its reference to the revealed perfections of God, and the constituent elements of every created mind. The distinction is found in the words of the first and great commandment, where the heart, the mind, and the strength, are separately named, and placed alike under the supreme law of perfect love. There is the Morality of good affections and desires, of wise and clear intelligence, and of that active zeal and energy, which carries into effect the decisions of the judgment, and the desires of the heart.

Moral Science, in this aspect, includes, first of all, a pure and lofty ideal of right affection and desire. It does not deal mainly and immediately, like human laws, with the outward conduct alone. Its empire has its foundations laid deep within the heart. It enjoins that the tree shall be good, that the inward feelings and affections of the heart shall be pure, in order that the life may be pure also. It is a mortal enemy to all counterfeit and unreal righteousness, which begins by cleansing the outside, while all within may be full of extortion and excess, defiled with impure lust and selfish passion. Its aim is to cast salt into that hidden fountain of the heart, from whence either sweet or bitter water must unceasingly flow.

This is the first and highest branch of morals, on its subjective side. For here it rises above the mists which settle on the sides of the mountain, or the diffracting influences that flow from the complex conditions of our human and earthly life. In this highest region it is easier than in the rest to trace out clearly the manifold results and corollaries of that great law of love, equally simple and sublime, which binds together the whole moral universe.

But if love has for its first and simplest character a desire for our own good, and the good of others, or delight in the contemplation of the Divine felicity, a second must immediately follow. It must search diligently into the nature of that good which we are bound to pursue, and of those evils from which, both for ourselves and all around us, we are bound to seek deliverance.

Ignorant and unwise affection defeats itself. It hinders

those whom it would help, and harms those whom it desires, perhaps eagerly and intensely, to succour and to bless. Human life and happiness are highly complicated things. There is wheel within wheel in that strange machinery, on which comfort and distress, ease and pain, moral progress and advancement, or decline and degradation, are found practically to depend. Mere wishes of benevolence may too often avail little, if there be complete ignorance or misconception of the true causes and elements of human well-being. It needs patient, humble, and persevering toil, self-knowledge, experience of life, and study of the Word of God, to trace out clearly those mysterious Nile-sources, 'that hide themselves from. careless eyes, from whence the wide and ample streams of human joy and sorrow perpetually flow. And hence one important division of Moral Science must consist in what Locke has made the title of a separate work,—"The Conduct of the Understanding." His remarks, however weighty, touch only the brink of a wide sea of thought, and dwell very slightly on those aspects of the subject, in which its ethical character, as one main branch of morals, most distinctly appears.

Here the old Socratic or Platonic inquiry—πότερον διδακτὸν ἡ ἀρετή, whether virtue is something that can be taught—finds its natural place. It links itself closely with the deepest problems, not only in natural metaphysics, but in revealed theology. Into these it would be premature to enter now. They could not possibly be dealt with in a few pages, in a manner worthy of their high importance. It may be enough here to observe that an obligation to seek the good and promote the welfare of others must involve the duty of earnest and continued effort to see clearly wherein that good and welfare really consist.

THE DIVISIONS OF MORAL SCIENCE. 107

The homely proverb, that "hell is paved with good intentions" scarcely deserves the wide acceptance it has found. We may say, with deeper truth, and a very solemn meaning, that bad intentions are the real pavement, on which is reared continually the dark superstructure of consummated vice and crime, and of open punishment, whether in this life or the life to come. What may be more truly said, in a figure, to be paved with good intentions is that limbo of vanity, where all things

Abortive, monstrous, and unkindly mixed,

all perverse, but well-meaning follies, due to careless haste, blind prejudice, or self-satisfied ignorance, wander up and down in darkness as their natural home.

A general desire for happiness is the natural instinct of all men. And some wish for the good of others, however feeble, is so plain a dictate of conscience, that even the selfish and vicious, who do not really feel it, constantly feign a virtue which, as they are well aware, they ought to possess, and the entire want of which must be fatal to their reputation among their fellow-men. But the interval is very wide between the mere instinct, or the feeble, careless, and irresolute desire, and that earnest effort to learn what is good and right, and then to practise and promote it, which is the true sign of a genuine and hearty benevolence.

A large part, then, of our moral duty must consist in the right conduct of the understanding. It must include the diligent, humble effort to learn, both in our own case and that of others, what is the true good which claims our zealous pursuit; what is the nature, and what are the elements of that happiness towards which we ought to aspire, and what are practically the best means by which it can be really attained.

A pretentious and boastful, but blind and ill-judging philanthropy is one of the chief moral dangers of the times in which we live. The grandeur of the object, the seeming nobility of the aim, its wide and comprehensive nature, too often veil the selfish pride, the rash folly, and in some cases the open impiety, which first sets up a mere figment in place of the great reality, and then falls down and worships blindly those wretched theories of human progress and political perfection, which its own fancy has conceived, or its own fingers have made.

True benevolence is humble and modest. It is suspicious of mere phrases. It puts no easy faith in hollow and shadowy counterfeits of that good, so high, noble, and excellent, which it keeps ever in view. It studies carefully the lessons of experience. It listens with reverence to whatever brings the credentials and reasonable marks of a Divine message. It does not despise the helps to human happiness, which spring from wise laws, social freedom, and the progress of men in those arts and sciences, which multiply the conveniences of social life. But it knows and feels that the main sources of peace and comfort must lie deeper, and be sought higher, than all these. They are land-springs that soon dry up and fail. They are plants too feeble to resist the storm, and may soon wither or be swept away in the whirlwinds of human passion. The men whose love to their fellow-men is most real, deep, and earnest, will distrust the cries of the passing hour, the uncertain voices of popular feeling, which vary from year to year, and almost from day to day. They will try to dig deep, and lay their foundation on solid rock, before they claim to become master-builders in so vast a work as the future and lasting regeneration of the human race. Or rather, they will feel that so great and noble an end must lie wholly beyond the range of their own feeble efforts. And while they labour faithfully, to the best of their power, and use whatever light they have attained, for the good of their fellows, they will look up with reverence for the promised working of a higher Power, and place their hopes of success in a goodness and wisdom far beyond their own.

But if high and lofty desire for the good of others be one main division of moral duty, and a second is Moral Wisdom, or growing acquaintance with the true nature of that good we should seek to promote and attain, the third is Moral Energy, or that zeal and earnestness which seeks ever to carry out in practice the inward convictions of the mind and heart.

Energy of will, it ought always to be clearly felt and owned, cannot alone constitute moral virtue. There are other conditions of right feeling, and true knowledge of duty, which must first be satisfied. Where there is neither goodness in the heart, nor wisdom in the understanding, mere activity, however restless and incessant, will only widen the range, intensify the amount, and deepen the shades, of moral guilt. To "do evil with both hands earnestly" can only place men at the widest possible remove from the summit of true perfection. The hero-worship which some, despairing of Christianity, would exalt into a kind of religious faith, is a poor and wretched substitute for Christian morality. Power is not goodness, and energy of will is not moral excellence. And yet, since man was made an active being, in the image of God, on whose ceaseless energy of wisdom and love all things depend, there is a natural fitness and beauty in human energy, even apart from the right or wrong direction it may assume. And when true benevolence is joined with earnest study of the proper means by which its desires may be attained, and with a hearty zeal for its actual attainment, then human virtue, amidst all its weakness and imperfection, becomes a lovely image and resemblance of that goodness which is infinite and divine.

III. Moral Science, however, besides a threefold division from the objects it embraces, and the natural faculties which it involves and requires, of feeling or affection, knowledge, and will, admits a further arrangement or partition of a still deeper kind. This depends on those moral capacities of obedience, transgression, and recovery, which are linked, mysteriously and inseparably, with the very constitution of a universe of moral agents. And here, once more, the whole science naturally resolves itself into three main divisions.

The first of these is Simple or Direct Ethics. Its primary law and formative principle is the love of being, as being. This unfolds itself, as we have seen, into Personal Morality, or Self-love, Social Morality, or the love of our neighbour, Divine Morals, or Piety and the Love of God. Or again, with reference to the human faculties, into Moral Aspiration, Wisdom, and Energy, or the ethics of the heart, the mind, and the strength. This last variety is well summed up in the precept: "Whatsoever thy hand findeth to do, do it with thy might;" and consists in that moral earnestness, which constitutes true heroism, when once it is devoted to the cause of truth and love. But in all these varieties we have first to consider moral duty in itself, or the love of being as being, without reference to the moral contrasts, which experience has revealed.

The second branch is compound, and might also be

styled Diagnostic or Judicial Ethics. It is that branch of the science which unfolds the law of right feeling and conduct towards all beings, when viewed as capable of immense moral diversity and contrast, from Divine perfection, or the highest forms of angelic or human excellence, to the worst and lowest degrees of folly, vice, malice and profaneness. Its double law is the love of all good, and the hatred of all evil. It includes the three kindred subjects, righteousness, judgment, and justice. First of all it prescribes a right and just condition of the affections, in contemplating all the actual or conceivable varieties of moral character, so as in every case to love and admire the good, but to condemn, despise, and hate, every form of moral evil. In the next place it enjoins a sound exercise of the judgment, so as to discern, not only the broad contrast of good and evil, but also the various degrees of virtue, moral goodness, and holy love, or of vice, depravity, lust and malice, and thus to guide the determinations of the will in all practical questions that may arise. In the third and last place it includes justice, or the actual, honest application and development of true convictions and right judgment, in reference to all those cases of moral contrast and diversity, which meet us, from time to time, in the experience of human life, and in the conflicts of a world, where light and darkness, good and evil, struggle and contend from age to age.

The third main division, which completes and crowns the whole outline of the science, is Therapeutic or Remedial Ethics. And here we have to begin by admitting a weighty and solemn truth—the past and present existence and wide prevalence of moral disease, which needs some powerful remedy for its effectual cure. Ethics, in this department, deals with moral beings, neither as upright and sinless, nor as hopelessly fallen and irreclaimably corrupt, but as those who need recovery, and may be recovered, from decay, weakness, and degradation, to uprightness and moral healthiness once more.

The parent truth, in this vast and large division of the science, is that great fact, which forms the basis of the whole economy of the Christian faith. Man, however far gone from original righteousness, or the standard of a pure and perfect love, is not therefore hopelessly fallen. A way of life, that leads upward, is still open before him. He may yet arise from the dust, loose himself from the bands of his neck, and stand upright, as a moral freeman once more, in the face of heaven. His sickness, however sore and grievous, and incurable by mere human or external remedies, admits of successful treatment by the skill and love of a Divine Physician. And the science of Remedial Ethics, viewed on the side of pure reason, involves these successive elements. First, the great duty of desire and longing for this moral restoration, both for ourselves and others, so far as it is needed, and its possibility has been revealed. Secondly, the duty of learning what are the means by which this great recovery to a new and higher moral life must first begin, and may still be carried on, with all the hindrances to its birth and later progress, and that provision of moral and spiritual agencies, by which these various hindrances may be overcome. Thirdly and lastly, the earnest practical adoption of all the means of recovery and moral restoration, which Divine love and wisdom may have provided for sinful men, and the ceaseless and persevering progress of the soul, from the miry clay of the slough of despond, towards that celestial city which is the dwelling-place of light, purity, and heavenly love.

These three kingdoms of Moral Science, for such they may fitly be called from their largeness and extent, are in other words the Ethics of Paradise, of the Law, and the Gospel. And each of them in turn admits a nine-fold subdivision, from the moral objects with which they deal, and the faculties of the moral being out of which they grow, and on which their existence depends.

Combining all three elements, we may form a comprehensive outline of the geography and whole domain of Ethics, which is, under another phrase, the Science of Ideal Humanity. It will then assume the following form.

A. Simple or Direct Ethics:

Its primal law and basis is Love, or genuine desire for the highest happiness or welfare of all sentient being, as being.

I. Personal Ethics = Pure and unselfish Self-love.

1. Self-love in affection and desire. This implies the aim at our own welfare in the whole range of our being.

   a. Bodily welfare or health; (a) as capable of sensation,—pleasure in contrast to pain; (β) as gifted with active powers, or Gymnastics in its proper sense; (γ) as capable of degrees of health and strength,—Dietetics and Hygienics; (δ) as capable of grace and beauty, or Cosmetics and Æsthetics.

   b. Welfare in respect of Intellectual Power. "Mens sana in corpore sano."

   c. Moral Welfare. The pursuit of moral goodness and spiritual excellence. The heathen maxim in its best sense,

Αἰὲν ἀριστεύειν καὶ ὑπείροχον ἔμμεναι ἄλλων:

or the higher law of the Apostle, "Covet earnestly the best gifts, and yet show I unto you a more excellent way."

2. Self-love, with reference to the powers of the mind, or Moral Wisdom.

a. Cultivation of sanitary science, wise self-preservation, the converse of the indirect suicide of vice and folly.

b. Mental Self-knowledge, wise discernment of special gifts leading to a right selection of the pursuits of life, and of work to be done.

c. Moral Self-knowledge. "Prove your own selves."

3. Self-love in Action, or Practical Culture, under the guidance of moral wisdom, of all natural and acquired gifts of body and mind.

II. Social Ethics = Philanthropy or General Benevolence.

1. Benevolence in affection and desire, applied to all the various relations of human life.

a. Conjugal or Married Love, in the relation of husband and wife.

b. Parental and Filial. The right laws of affection in the fourfold relation of father, mother, son, daughter.

c. Fraternal, or the laws of right affection in brothers and sisters.

d. Domestic. Relations of master, mistress, and servant.

e. Local. The duties of neighbourhood.

f. Friendship. Bonds and laws of special intimacy and affection.

g. Civic Relations. Magistrate and citizen.

h. National. Relations of sovereign and subject.

i. Legislative. Varieties of social rank and constitutional privilege.

j. International. The moral laws of right and benevolent feeling with regard to other races, and other civil communities.

2. Benevolence of the Mind, or right study of the means of promoting the welfare of others.

a. The Relations of Sex. *b.* Of Age. *c.* Domestic Economy. *d.* Civil Economy. *e.* Political Economy. *f.* National Economy.

3. Beneficence = Social Morality in practice and outward result, with reference to all the duties and relations of daily life.

III. Divine Ethics = Piety, or the Love of God, including

1. The Religion of the Heart; Adoration, gratitude, obedience.

2. The Religion of the Mind. Divine contemplation, and the use of all means for growth in knowledge of Divine things.

3. The Religion of the Strength, or pious and holy affections, and a true knowledge of God, applied faithfully in the active and outward duties of a religious life.

B. Compound or Judicial Ethics.

Its primal law and formative principle is Righteousness, or to feel and judge of all moral agents, and act towards them, according to their various characters, good or evil, sinless or fallen, penitent or impenitent, proud or humble, advancing or declining in their moral state. And here we have, as before—

I. Personal Ethics, including

   a. Self-condemnation and self-hatred, as evil.

   b. Self-respect and self-honour, so far as upright, sincere, and faithful.

   c. Repentance and humility.

   d. True and right judgment on the personal state, whether of moral progress or moral decline.

   e. Earnest aspiration after full victory over evil, and a perfect restoration to moral perfection.

II. Social Righteousness, including

   a. Love and esteem of the good for their goodness.

   b. Hatred and contempt of the morally evil as evil.

   c. Moral discernment of the true character of men.

   d. Practical righteousness, in praise and blame, reward, censure and punishment.

III. Divine Righteousness.

This includes the whole doctrine of the revealed law of God, with its disclosure of the Divine holiness, its threatenings and promises, in their connection with the Divine attributes, and the application of its truths to the hearts and consciences of men.

C. Remedial Ethics = the Morality of the Gospel, viewed as a Divine provision of mercy for the healing of moral disease. Here Ethical Science rises into a still higher field. It becomes, in the truest sense of the term, Moral Theology, and links itself with all the promises of the Gospel and the hope of the life to come. It is transfigured, and loses its earthly features in a higher light that beams down upon it from heaven.

This outline, however brief and imperfect, may help to give some conception of the width and variety of those fields of high and sacred thought, which lie open before the general students of Moral Science. To unfold them all, as their grandeur and rich variety demand, would far exceed the work of a whole lifetime. But to open out, in any degree, their hidden treasures, and to throw clearer light on any of their hard and difficult problems, is an object worthy the ambition of every Christian heart. The moral teacher, who is at all faithful to so high a trust, may have to encounter misconception from the careless, or ridicule and obloquy from moral triflers, in an age prone to self-flattery, and far too impatient of calm and serious thought. But these difficulties are only like dust in the balance, and utterly to be despised, when compared with the privilege of tracing out these pathways of high and holy thought, which lead the spirit upward to the footstool of God's throne, there to lose itself in the light which flows from His presence, who is uncreated, infinite, and perfect goodness.

This lesson of the true dignity of Moral Science, as superior to the noblest triumphs of human skill in the fields of nature, is taught most clearly in the words of the patriarch, in the oldest book of inspired and sacred poetry. A modern paraphrase of this voice of early antiquity, which is full of such deep meaning, may bring this subject to a fitting close.

Man, we are told by the tempted patriarch, is diligent in searching out the hidden treasures of the earth. These are all of them objects of his assiduous and successful pursuit; the veins of silver, the place of sapphires, and the dust of gold. "He searcheth out all perfect things, the stones of darkness, and the shadow of death." He tills the earth, and reaps its plentiful harvests, so that "out of it there cometh bread." He turns up its soil, and brings to light the gems, sparkling like fire, that were buried in its bosom. There are paths which no fowl knoweth, and the eye of the vulture cannot see, and which the wild beasts of prey, roving over the forest and wilderness, have never trodden. These are the pathways of Man, when he sinks the shaft of his mine, puts his hand upon the rocks, and pierces the sides or depths of the mountain, that he may bring to light all its stores of hidden treasure. But amidst all his labours, however prosperous, one thing still eludes his search, and seems to baffle all his industry, and the pride of his science. "Where shall wisdom be found? Where is the place of understanding?" How shall he attain that moral insight, that true ideal of human action, speech, and thought, whereby he may guide his steps aright in the weary journey of this mortal life, in which his days seem often to be "few and evil"? How may he escape the follies of childhood, and the snares of youth; and be freed from the misery of a life wasted without object or aim, cursed with vanity and vexation of spirit in every moment of its swift progress to the darkness and mystery of the grave?

Here is a hard problem. The lapidary or the miner, in the days of the patriarch of the eastern desert, could not solve it. And it remains wholly unsolved by theories of evolution, or the researches and discoveries of physical philosophy to the present hour. We mine deeper than ever into the bowels of the earth, and bring up fossils, buried long ages ago, and eloquent of the past, but "the depth saith, It is not in me." We dredge the ocean with industry and consummate skill, and from those profound abysses bring up new materials for the growth of science, but "the sea saith, It is not in me." We heap together wealth in its most various forms, and all the costly treasures of art; but these do not yield that higher gift of which we are in utmost need. "It cannot be gotten for gold, nor can silver be weighed for the price of it." It cannot be bought in exchange for the most precious jewels, for coral or pearl, onyx or sapphire. Nay, even if all the costly things of earth are summed together, and formed into one coronet or diadem of exquisite beauty, they will not buy this inestimable blessing.

"Whence then cometh wisdom?" How shall we gain that knowledge of highest worth, which may guide us through the dangers of life's journey, and bring us to some haven of peace and rest? When Science fails, and all nature, being eagerly questioned, gives up the task of answering in despair, a higher voice makes answer. He who is the Creator of this outer world, with its complex powers and mysterious treasures, is also the God of the spirits of all flesh, the Author and Parent of our moral being. He makes a decree for the rain, and a way for the lightning and the thunder. He fixes the circuits of the winds, and weighs the waters by measure. But this wonderful Geometer is also the Father of lights, the Parent and Guide of the spirits of men. He has not left us to grope blindly among His lower works for that peace and hope, which can never be found in these alone. He does not condemn the children of men to thirst for a wisdom which will never be given, and for light on the wants, duties, and destinies of the soul, which can never be attained. The deeper mines of moral truth are all in His sacred keeping. The precious jewels of truth, wisdom, and love, that may enrich the spirit for ever, are subject to His secret control. Let men only awake to their true dignity, as the creatures of God, dependent on His teaching, and their quest for moral truth and insight shall not be in vain. "When he made a decree for the rain, a way for the lightning and the thunder, then did He see it and declare it, He prepared it, and searched it out. And unto man He said,—Behold the fear of the Lord, that is wisdom, and to depart from evil is understanding."

In this sublime apostrophe of the patriarch, we have a key to the true relation between all physical science, and that higher field of thought which deals with moral right and wrong, sin and duty, the lessons of social Ethics and of religious piety and faith. These lower pursuits, in their own place and order, have a great value. But no abundance in the outward arts of life can ever secure human happiness, while the passions of discontent and pride, envy and malice, are reigning within the hearts of men. To depart from evil alone constitutes true understanding. And this victory over moral evil, whether in the individual or in nations, can never be really and permanently gained, unless moral instincts are confirmed and deepened by the presence of religious faith. The two lessons of the patriarch's message always go together. They have been married by the Supremely Wise, and no human power can effect their divorce and separation. Genuine piety, or the fear of God, must lead the way; and a high social morality, which discerns evil clearly, and then departs from it with a hearty aversion, is the happy sequel that must ever follow.

This lesson of ancient wisdom is receiving hourly new illustrations in the times in which we now live. A vain and misleading philosophy may extol the power and celebrate the triumphs of intellectual genius, and degrade all moral truth into a meagre compendium of worthless and almost unprofitable truisms. But in such comparisons

THE DIVISIONS OF MORAL SCIENCE. 121

we have only a proof that its own eyes are dim with age, so that it cannot see. The grand stars of the moral firmament are settled in the heavens, and shine on for ever. The wanderer, here on earth, may lose sight of them amidst the glare of meaner and earthly lights, of passing and transient meteors, or they may be hidden from his view by the mists and fogs that thicken around his steps. But when once he looks upward with patience, in calm thought, their holy light will dawn upon him once more. At first it may seem very dim and feeble. But it is the sign of coming daybreak, the earnest and pledge of clearer insight and vision that will presently be given. The higher and nobler the science, the later and the longest delayed we may expect its full development to be. "That is not first which is spiritual, but that which is natural, and afterward that which is spiritual." Even Physical Science is still only in its childhood, and an immense progress, with deeper insight into laws of nature still unknown, seems to be its heritage in coming generations. But whenever the long-promised moral daybreak of our world shall arrive, all previous discoveries in the field of physics, however large in themselves, will seem almost mean and trivial, when compared with the rich harvests of pure and lofty truth to be reaped in this higher and nobler field. For Moral Science is nothing less than the grand ideal of redeemed humanity, of which the seeds and germs are planted already, but their perfect growth and expansion can only be when the redemption is fully come. It deals with that high and pure standard of human thought, speech, and action, in which heaven stoops down to the succour of human weakness, and all the relations of earth are raised and transfigured, and lose themselves in the light of heaven. The ideal is so bright and beautiful that our eyes are soon dazzled, when we attempt to unfold its features by reasoning alone. It can only be seen and known, loved and admired, as it ought to be, when we rise above the diffracted and broken earnests of mere human virtue, and gaze on the true standard of humanity, where all moral perfection is summed up and condensed, without limit or measure, in the person of the Incarnate Son of God.

LECTURE VI.

MORALS IN RELATION TO PHYSICAL SCIENCE.

IN the world of scientific thought, no less than in the sphere of politics, questions of boundary and of jurisdiction are of high importance. Of the former truth our own country has had recent and painful experience. The carelessness or ignorance of negotiators may breed national disputes, even on matters comparatively insignificant, which become a legacy of bitterness and strife to successive generations. And if our country has had recent experience of the evils arising from boundaries ill and ambiguously defined, its earlier history supplies examples, both in its wars with France and Scotland, of the mischief arising from unsettled rights, or claims of jurisdiction and feudal superiority, even on a larger scale. It is a matter, then, of no slight importance, to decide on the true relation between Physical and Moral Science, to know where the boundary lies, and whether there is a just claim, on either side, to superiority and jurisdiction.

In the middle ages, and during the reign of the schoolmen, there were serious and mischievous encroachments, by theological moralists, on the domain of Physics. Ingenious *à priori* theories of creation and its laws, such as in their imperfect fancy they ought to be, replaced the use of diligent and patient inquiry into the facts themselves. The mischief revealed itself in two forms—a hasty interpretation of Scripture, and a blind reliance on the authority of Aristotle. And of these the latter was not the less frequent, while it is less easy to palliate or explain. When Scheiner announced to his Provincial his own discovery of spots in the sun, he received this answer: "I have read Aristotle many times from beginning to end, and can assure you that I have nowhere found in him anything similar to what you mention. Go, therefore, my son, tranquillize yourself. Be assured that what you take for spots in the sun are the faults of your glasses or your eyes."

These faults of theologians and moralists in former days seem to have caused a strong reaction in our own. The turn of Physics has come, to make bold and extensive encroachments on Moral Science. We may notice two different degrees or stages in the assumptions of superiority which some of its disciples have latterly made.

The first and most complete may be seen in the attempt to destroy Morals, properly so called, altogether. Men may busy themselves so eagerly among laws and changes of a physical and material kind, that their conscience falls asleep, and they grow insensible to the human elements of their own being. In this case they end by denying free-will and responsibility. All changes are resolved into some mysterious and fatal necessity. Conscience itself becomes an acquired and secondary instinct of the animal life, or a strange resultant from the combination of material forces, condensed by the nervous system in some unknown way. Such a theory, it is plain, puts an end to Moral Science altogether. The name may possibly be retained for a time, but the substance is gone. What inherits the name of Ethics is merely a branch of physiology, or some transcendental formula of dynamics, which only awaits its solution by some "competent intelligence," some profound and subtle analyst in a future age.

The other form of usurpation or rivalry is more limited and moderate in its hostility. It is willing to allow that Morals have a sphere of their own, and that man is, in some real sense of the word, a spiritual being. But it claims that moral elements shall be excluded wholly from the domain of Physics. It protests against the doctrine of final causes, as a gratuitous disturbance in our search for the genuine laws of nature. It refuses to see, in the frame of nature, in the structure of plants and animals, any real marks of design, and would thus exclude the notion of a Moral Governor and Designer, if not from the whole field of human thought, at least from the range of the material universe.

It becomes, then, highly necessary for the student of Morals, in the presence of such claims and theories as these, to understand clearly, and hold firmly, the true relation between these great kingdoms of human thought. Their entire separation is unnatural and even impossible. It contradicts and reverses that growing tendency to unity, which marks all the recent discoveries of physical philosophers themselves.

The relation, then, of Ethics to Physics may be conveniently viewed under these three heads; (1) Resemblance or Analogy; (2) Difference or Contrast, and (3) Supremacy and Inalienable Jurisdiction.

There is, first of all, between Ethical and Physical Science, a strong resemblance or analogy.

What are the main elements, which enter into all the various and complex researches of Physical Science? There are four main facts or principles, on which, when we examine it closely, the whole superstructure will be seen to depend.

And first, there is a Fundamental Fact of Experience, or the existence of outward objects, which we can see, feel, handle and observe. Thus we gain the conception of Space and a Material World. The materialism, which denies mind altogether, has sometimes, by reaction, given birth to a mystic idealism, which resolves matter into states of thought, and denies the existence of an outward world. Such a theory, if the instincts of common sense did not rebel against it, and set it aside virtually even with its own disciples, would be fatal to the existence of all Physical Science.

A second main element in all Physics is the recognition of fixed and necessary laws or relations among these outward objects, such as the truths of Arithmetic and Geometry. Their distinctive nature is that we not only receive them as true, but we cannot conceive them to be reversed. Their opposite is recognized as a self-contradiction.

A third element, distinct from the last, is the presence of laws which are actual, but not conceived to be necessary, and which are found to link past, present, and future changes together. Their discovery, so as to infer from the present state of the visible world an actual past and an expected future, is the main object of all physical research. These laws do not result from *à priori* reasoning. Their existence has to be determined by careful deduction from observed facts. It is not an inference of pure reason, but depends on actual experience and observation alone.

Lastly, in all physical science there are actual conditions, whether constants or variables, which constitute the materials to which laws have reference, and without which those laws could not even exist. These are almost innumerable, more numerous than the atoms of which the universe is composed. On their mutual relations and incessant changes all the phenomena of Physics depend.

Now Moral Science has the closest resemblance to Physics in these its four main constituent elements. Its fundamental fact is the Conscience, or in other words the existence of Moral Agents, gifted with the power of choice, capable of knowing good and evil, in whom feelings of moral praise or blame spontaneously arise. Its first postulate, answering to that of Physics, is the reality of a moral universe. Next, there are necessary laws and relations, discernible by the conscience, and the opposite of which, when examined, give rise to a feeling of self-contradiction and inconceivability. Such are the truths and first principles of Pure Ethics, the hatefulness of treachery and malice, the beauty and excellence of genuine love.

Still further, in Morals as in Physics, there are laws which are found to be actual, but are not felt to be necessary, so that a different constitution of things may readily be conceived. These are conclusions from actual experience with regard to the faculties of man and his moral capacities, and the special relationships and resulting obligations of human life. Last of all, a completed moral science involves a knowledge of the actual state of the moral universe, at least of that part which is accessible to us, the race of mankind. It includes a wide range of experience, and a full acquaintance with the varieties of human character, and the countless forms of probation, and moral advancement or decay.

In these main features Morals and Physics have the closest resemblance. Both alike include the phenomenal and the real, the actual and the necessary, changing circumstances and invariable laws, facts of experience and deductions of reason. It is on right apprehension of these, in their essential distinctness, and their inseparable union, that the growth and reality of either science depends.

But the relation of Ethics to Physical Science includes also some important features of difference and contrast.

The first contrast, which must strike every candid and thoughtful mind, is in their dignity and importance. Physics, in its narrower sense, deals with lifeless matter. Botany and Zoology rise higher, and deal with living things, plants and animals, of which the highest powers are animal instinct alone. Moral science deals with persons, not with things. And the persons, who form its object, are not merely sentient, but reasonable beings. They are capable of knowing themselves, and each other, and of rising to the knowledge and love of the Father of their spirits, the Creator of the whole universe. Man is higher than mere animals or lifeless matter; and the science which deals with human thoughts and duties, hopes and fears, must be nobler than even the highest departments of philosophy.

There is a further contrast of necessity and freedom. Matter is subject to a law of physical compulsion. It has no choice but to obey. Obedience here is perfect, and deviation impossible. But Moral Beings have a power of choice. The law under which they are placed is one of duty and right, not of fatal necessity. It is one which ought to be kept, but may be broken. The Actual and the Ideal here diverge and part from each other. And as the powers entrusted to such agents are of a higher kind, so in their case there is the possibility of a deeper fall. Experience proves how wide may be the melancholy interval between the law of duty and the actual conduct of mankind.

There is a difference also of colour and tone. For Physics, in a figurative sense of the word, may not unfitly be styled a colourless science. It dwells in a region of dry light. Its truths have only a remote and indirect relation to the feelings of the heart. But Morals are a science to which that name is only sparingly applied, because its truths are more than objects of intellectual perception alone. They are vocal with strong emotion, and awaken deep echoes in the hearts of men. They stir the depths of the spirit with feelings of sympathy and antipathy, of remorse and self-approval, of praise and blame, and open out hidden fountains of pity and gratitude, of enthusiastic admiration and earnest love.

In the practical aim of the two kinds of knowledge there is also an important difference. Physics increase the power of man over nature, and thus help to secure his comfort in all economy of his outward life. But the deepest wants of the human heart lie beyond its range. Its discoveries, however important in their own sphere, leave these unsatisfied. It cannot tell men why life has been given them, and to what uses it should be applied. It solves none of those doubts and questionings which arise in their conscience, in the pauses of business, or in the prospect of death and the grave. It throws no light on the dark problem of moral evil, on the secret of human guilt, depravity, and sorrow, or on the nature and the symptoms of a disease so melancholy, and the possibility of its cure.

There is another important contrast between Physics and Morals in the conditions on which their progress depends. In Physics these are intellectual only. All that is needed is to bring to the inquiry a clear understanding, and the power of lucid and continuous thought. But moral insight requires moral, no less than intellectual conditions, in those who desire to attain it. There must be singleness of purpose, a real desire to learn the truth, and a willingness to obey that truth, so far as knowledge of it has been attained. The sentence among the sayings of Ali is both simple and profoundly true: "Knowledge calleth out to Practice. If it answer, well. If not, knowledge goeth away." The same principle has the sanction of a far higher authority. "If thine eye be single, thy whole body shall be full of light." "Blessed are the pure in heart, for they shall see God." To attain fuller and clearer insight into moral truth, one first requisite is practical obedience to the measure of truth already attained.

But the relation of Morals to Physical Science is not merely one of fundamental harmony and partial contrast. It includes further a lawful claim to supremacy and jurisdiction.

These two main kingdoms of thought, though widely distinct, and marked by some strong features of separation and contrast, are by no means wholly independent. On close inquiry, they will be found to have bonds of close and inseparable union. Whenever ethical truth has to be applied to the actual circumstances of our human, mortal life, Physical Science will be found to have an important bearing on all the details of moral duty. But

Moral Science claims jurisdiction and supremacy over all departments of Physical Study, on the clear and simple ground that these last form one main field of human action. They must therefore abide under the dominion of the great laws of morals, alike in the first opening of physical research, in its later progress, and its final application. Moral ideas are not excluded from the field of Physics, as some in these days have very erroneously conceived. On the contrary, they form one essential and inseparable constituent in all its problems, when once we look below the surface, and examine the deep foundations on which they rest.

To gain such a knowledge of natural laws, that from observation of present phenomena we may be able to trace out past, and predict future changes, is the grand problem of physical research in reference to the outward and material universe. But the lessons implied in this hard problem seem not seldom to be overlooked and forgotten, even by scientific men. Let us suppose the complex and transcendental function to be ascertained, from which future analysts of competent intelligence, whether by integration or infinite series, are to deduce from the present state of our earth the structure of a primitive nebula, countless ages ago, and then to predict with certainty, not only the eclipses and meteoric showers, but the wind and the weather, and perhaps even the flights of locusts and birds of passage, in distant ages to come. Let us lay aside for the present the natural inquiry, whether animal instincts, and still more the human will, may not disturb the meshes of that physical constraint and necessity, in which these theorists thrive, like the Philistines, to bind down the mighty forces of the spirit of man, and make it work and grind in chains in a dark prisonhouse of destiny. Still, when this mighty formula is framed, on the solution of which the vast problem is to depend, it must include the three coordinates of every atom of matter, or of any physical medium more subtle than common matter, in the whole created universe. These are the conditions of the problem, the indeterminates that need to be determined, and on which the solution must depend. The time enters only as one indeterminate in addition to all the rest. Now Physical Science proposes simply this problem, from the known present to infer the changes of past time, and predict those of the distant future. Assuming all the other data to be ascertained, it would express the state of all the parts of the outward universe as functions of the time alone. Thus its utmost triumph is to get rid of one variable out of a number three times greater than the number of all the atoms of the material universe—a number amounting to trillions of trillions, and practically infinite. All these countless conditions of its own vast problem physical laws cannot in the least explain. For the laws might be unaltered, and these elements, on which their application depends, might be infinitely varied. Let us assume the existence of a thousand bodies only, and that their motions, through successive ages, are determined by their mutual gravitation alone. Still we might multiply a million into itself a million times, and not exhaust the conceivable varieties of the first position, by which, in virtue of the law, every later position would be determined. How, then, can the actual values of these positions be explained? By transferring our thoughts, either to the remote past or distant future, we do not lessen their number: it remains unchanged. They can be accounted for in one way alone. From the laws we must rise to the Lawgiver. We must accept the ideas of creation, and of an Intelligent Cause; a Divine Person, who is capable not only of establishing laws of nature, but of choosing out those elements and conditions for their activity, which satisfy objects of a higher kind, and subserve the purposes of a mighty scheme of universal providence.

It has been affirmed lately, by a skilful and popular experimentalist, that "the scientific mind can find no repose in the mere registration of sequence in nature. The further question intrudes with resistless might—whence comes the sequence? What is it that binds the consequent with the antecedent in nature? The truly scientific intellect can never attain rest until it reaches the forces by which the observed succession is produced. It was thus with Torricelli. It was thus with Newton. It is thus preeminently with the real scientific man of the present day. He knows further that the succession, besides being permanent, is *necessary*. Not until this relation between forces and phenomena has been established, is the law of reason rendered concentric to the law of nature; and not until this is effected does the mind of the scientific philosopher rest in peace."

These remarks are true, so far as they affirm the transition from mere phenomena to causes, forces, or powers, to be the genuine instinct and tendency of a philosophic mind. This is in reality the exact antithesis of Modern Positivism, which places force and power among those hurtful metaphysical abstractions, which mark the immature youth or childhood of science, and affects to limit all sound research to the registration of phenomena alone. But when they affirm that necessity enters into the true scientific conception of every law of nature, they fall into an error no less glaring and conspicuous than the positive doctrine which they condemn. It is hard to explain the process of thought, by which a view is ascribed to the first of physical discoverers, and enforced by the authority of his name, which he has deliberately rejected, and has even devoted the last pages of his immortal work to its brief and decisive refutation. This doctrine of physical necessity he places in the same rank with the theory of vortices, and treats it as still more opposed to sound, philosophical reason. He writes in his closing Scholium as follows :—

"Elegantissima haecce Solis, planetarum et cometarum compages nonnisi consilio et dominio Entis intelligentis et potentis oriri potuit. Hic omnia regat, non ut animi mundi, sed ut universorum dominus. Hunc cognoscimus per sapientissimas et optimas rerum structuras, et causas finales, et admiramur ob perfectiones ; veneramur autem et colimus ob dominium. A cæcâ necessitate metaphysicâ quae utique eadem est semper et ubique, nulla oritur rerum variatio. Tota rerum conditarum pro locis et temporibus diversitas, ab ideis et voluntate Entis necessario existentis solummodo oriri potuit."

But the falsehood of the view does not rest solely on the high authority of Newton, to whom an appeal in its favour has been so strangely and rashly made. It results immediately from careful reflection on the nature of known physical laws themselves. The only necessity which a really scientific mind can recognize in them is a hypothetical necessity, or a certain connection between the law, assumed as real, and acting alone, and motions or positions that will follow. Assume such forces to operate, and that they operate undisturbed and unmingled, and motions or changes of such and such a kind will assuredly result.

But there is here no answer at all to the question why these laws, and not others, actually exist, since others are equally calculable and conceivable. There is no proof, and hardly a presumption, that these laws, so far as they are known, may not be hourly modified by other physical laws, which are still unknown; or intersected and varied, constantly or periodically, by the action of other laws, dependent on spiritual agency, of a different and higher kind. Still further, the existence of the law, when admitted, furnishes no key to the special elements, almost infinite in number and complexity, on which its whole operation depends. These countless indeterminates, amidst possibilities still more countless, can be accounted for only by the choice of an Intelligent Creator, who has selected, out of countless possibilities physically indifferent, those which fulfil the secret aims of His moral government of an intelligent universe. By this key they may be perfectly explained, and by this alone.

Moral ideas, then, are not excluded from the field of Natural Philosophy, but, as Newton felt and taught, are its highest and crowning portion. The utmost attainment of mere physical research is to link the details of the present, by some law of connection, with the past and the future. But if we would answer the question, why the innumerable conditions which experience reveals are such as it reveals, and no other, when our reason clearly perceives that the possible alternatives, which have no real existence, are still more innumerable, we must rise beyond mere Physics into a higher field; and recognize with reverence the supreme will and choice of the Divine Creator, " to discourse of whom from the appearances of things does certainly belong to Natural Philosophy." The world must have been framed for a higher end than to give room for the evershifting and aimless motion of lifeless atoms. It was formed, by Divine Wisdom, to be inherited by creatures capable of worship, obedience, and love.

Physical studies, again, in their very outset, need to be guided and animated by a moral purpose.

Why am I to seek a fuller knowledge of the world around me? Is it to be only the satisfaction of a blind impulse, like that of the butterfly that passes idly from flower to flower, or the tiger that seizes fiercely on its prey? Men of science claim instinctively for their pursuits a higher character, and place them above the level of mere selfish trading or luxurious self-indulgence. Truth is the prize at which they aim. They long to pierce below the surface, and to gain a fuller and clearer insight than others around them, or even their own predecessors have attained, into the reality of things. Their ambition is to gain some κτῆμα ἐς ἀεί, a lasting inheritance of truth and knowledge for ages to come.

Such a quest of truth in the field of Physics is right and good in itself. But there are three conditions or moral laws it must obey. First, it should be the love of truth as truth, for its own sake, and not as a mere engine of power to astonish the vulgar, or to secure some mercenary object. They must use all their attainments in the spirit of love, and adopt the well-known prayer of Bacon, that the Father of lights may enrich their fellows with some new benefits through their labours. And it must be a love of truth in its due order and proportion. It is not only an intellectual anomaly, but a moral evil to be condemned and deplored, when truths of secondary importance are sought with eager diligence; while truths far higher, the grand lessons of moral duty, and the mysteries of religious faith and hope, are passed by with careless indifference, or even rejected with open scorn, and thrust contemptuously away.

Not only the first steps in physical research, but its whole later progress, is subject to the laws of moral obligation. All the treasures it pours at the feet of its disciples, like mere material wealth, need to submit to a double rule on which their right use depends. That rule is one of distribution and consecration. Physical knowledge, it is plain, ought never to be viewed as a private monopoly, for the benefit and honour of a few philosophers alone. And this obligation, too often overlooked or denied in former ages, is usually recognized in our own day. But there is another rule of a still higher kind, which is often disobeyed. The more we learn of the secrets of nature, the fuller and larger is the tribute of honour and reverence we are bound to pay to the God of nature, the Author of that wonderful and mysterious universe, whose hidden secrets He permits us to explore.

It was said once, in Newton's days, that "an undevout astronomer is mad." An opposite saying of M. Comte, in late years, has gained a mournful celebrity, that the heavens now reveal no glory but that of Copernicus, Kepler, Newton, Herschel, Laplace, and the other great discoverers in astronomy. That science, if this were true, would have become a fallen star. It would have dropped, like a meteor, from its place in the moral firmament. But such a hateful degradation of a noble pursuit, however widely it may spread for a little moment, is unnatural and monstrous. *Ilœ nubeculæ transibunt.* Soon or late the moral law, neglected for a moment by some physical students, will resume its just authority, and the words of the Psalmist become the keynote of all physical research once more: "In His hand are the deep places of the earth, and the strength of the hills is his also. The sea is His, and He made it, and His hand prepared the dry land. O come! let us worship and fall down, and kneel before the Lord our Maker!" "How excellent, O Lord, is Thy name in all the earth, who hast set Thy glory above the heavens!"

Once more, all physical researches, when successfully completed, need to receive a moral application. This truth is beautifully expressed by Lord Bacon in these well known words:—

"Wherefore, in the opening of our work, we pour out most humble and ardent prayers to God the Father, God the Word, and God the Spirit, that mindful of the cares of the human race and of the pilgrimage of that life, in which we spend days few and evil, they would deign, through our hands, to enrich with new benefits the human family. And this moreover we humbly pray, that human things may not injure the divine, and that from the opening of the gates of sense, and a greater kindling of natural light, there may arise in our minds no unbelief and darkness towards divine mysteries. But that rather, from a pure understanding, purged from phantasies and vanity, and not the less remaining wholly subject to the divine oracles, the things of faith may be given to faith. Finally, that the poison of knowledge, infused by the serpent, whereby the human mind is puffed up and inflated, being put away, we may be wise with sobriety, and cultivate the truth in love."

There are many different ways in which Physics, rightly pursued, may fulfil a high moral purpose, and own the superiority and minister to the progress of ethical truth. They serve to enlarge and ennoble our conceptious of the vastness of creation. They reveal, in every part, the wide dominion and supremacy of law, and thus convey a lesson of deep significance, when we transfer it to the higher field of human thought and action. They exhibit, in a striking picture, in all the celestial motions, and the unvarying constancy with which the physical laws of nature are obeyed, the beauty and excellency, among moral agents, of a perfect and faultless obedience to the supreme standard of right and truth. They disclose also, the further and deeper we search out the secrets of nature, an immense want and void in all mere physical theories of the universe, which can only be filled up and supplied, when we pass beyond them, and see in them the outer porch to a glorious temple of moral and spiritual truth. And lastly, when we gaze upon the fields of nature with patient thought, and trace out the structure and manifold harmonies of the material creation, we may see them stored with wonderful and various analogies, framed by the Supreme Architect and Governor, in order to raise our dull thoughts, through physical emblems and earthly shadows, to the clearer discernment of the nobler truths and deeper mysteries of the spiritual world.

The modern theory, which would isolate Physical Science wholly from moral truth, and form the visible universe into a gigantic preserve, sacred from all intrusion of direct spiritual agency, where blind destiny alone is to rule, undisturbed by any subordination to a moral purpose, and condemned to roll for ever the stone of Sisyphus, without ministering to the moral government of responsible creatures—like the old hypothesis of Cartesian vortices—"is pressed with many difficulties." It contradicts and reverses that tendency to unity which is one main feature in the whole course of recent scientific discovery.

It leaves unexplained, and wholly inexplicable, nearly all the constituent elements of every physical problem, those constants or conditions, without which no physical law can exist, and on which its vitality wholly depends. It leaves the choice of the laws themselves, out of many others equally conceivable, an enigma of which no solution can be found. It degrades the physical universe into a wide, vast chaos of unmeaning changes, a machine with no work to do, and by which no real work is ever done, a maze without plan, purpose, or object; where mighty forces are changing their shape every hour, and still, in every shape alike, are subject to the curse of utter vanity; where countless suns, and planets, and systems, and forms of transient, perishable life, find their true description from first to last in the mist out of which they are supposed to have been born,—

A thousand wreaths of dangling water smoke,
That, like a broken purpose, waste in air.

How far nobler and higher is that view of the relation between Physics and Morals, which genuine science and true Christian faith equally serve to reveal! "All creatures strive to ascend, and ascend in their striving." All nature points and looks upward. All outward laws subserve a moral purpose, and are wisely selected to constitute a worthy theatre for a grand scheme of moral government. Even when man turns away his eyes from heaven, and refuses to look upward, he may find, in the wide and vast ocean of physical change that lies beneath him, the reflections and images of heavenly things. The senses of the body represent to us the higher moral perceptions and capacities of the human soul. The starry spaces, and the unfathomed depths of earth and ocean, represent equally the vast range of unexplored mysteries in the spiritual universe. But when we seek to rest in these shadows, and mistake them for the substance to which they are Divine way-marks, they disappoint us, and elude our grasp. We seek a portion in them, but cannot find it. The whole range of Physical Science must ever remain, to the immortal spirit of man, a desolate wilderness, a land of darkness, shadows, and death; until we own in it the work of an all-wise Creator, the destined theatre for a wide dominion of perfect goodness. We may then look forward in hope to a time when order shall come out of seeming confusion, and light out of darkness; and when all the laws of nature, and all the perplexing changes of providence, shall be seen to be the appointed handmaids to one grand, eternal counsel of Creative Wisdom and Redeeming Love.

LECTURE VII.

MORALS AND POLITICAL ECONOMY.

THE relation of Ethics to Political Economy, even more than to simple Physics, is a subject of deep and growing practical importance in the present day. The want of clear and just ideas on their close connection, with the inseparable dependence of true economical science on the grand lessons and laws of Social Morality, has led already to very serious evils. How these may be resisted and overcome is one of those weighty problems, which claims for its solution the earnest effort of every Christian mind and heart. No remedy can be found for them, unless the basis be laid in sounder and deeper views of the whole question than have usually prevailed.

Political Economy is an ambiguous phrase. The sense in which it has been commonly used diverges sensibly from its proper and natural meaning. For Economy, as Sismondi has well observed, means by its derivation "the law of the house." The compound phrase, then, should naturally denote principles or laws to regulate that larger household, which forms a human state or political society. But the science usually intended by the phrase has a more limited range, and might perhaps be more fitly styled, "Natural Polity." It is that branch of Political Science, which deals with outward nature, and the material objects that surround us, and teaches how these may be best adapted and applied to the social wants of mankind.

The looseness and inexactness of the name has been attended with no slight degree of uncertainty and variety in the definitions of this modern science. Sir J. Stuart says that "its principal object is to secure a certain measure of subsistence for all the inhabitants of a country, to obviate every circumstance which may render it precarious, and to provide every thing necessary for supplying the wants of society." M. Storch says that "it is the science of the natural laws which determine the prosperity of nations, that is to say, their wealth and their civilisation." M. Sismondi defines its object to be "the physical welfare of man, so far as it can be the work of government." M. Say remarks that "it combines the results of our observations on the nature and functions of the different parts of the social body." Mr M'Culloch defines it "as the science of values, or of the laws which regulate the production, accumulation, distribution, and consumption of those articles or products that are necessarily useful or agreeable to man, and possess exchangeable value." But he adds that "its object is to point out those means by which the industry of man may be rendered most productive of wealth, to ascertain the circumstances most favourable to its accumulation, the proportions in which it is divided, and the mode in which it may be most advantageously consumed."

Mr Senior objects to some of these definitions, that they are too ambitious, and include a range which "far exceeds the bounds of a single treatise, or the powers of any single mind. The questions to what extent and under what circumstances the possession of wealth is beneficial or injurious, what distribution of wealth is most desirable, &c., are of great interest and difficulty; but no more form part of Political Economy, in the sense in which we use the term, than navigation forms part of the science of Astronomy." "The subject of legislation," he observes, "is not wealth, but human welfare. On the other hand, Political Economy treats not of happiness, but wealth. Its premises consist of few general propositions, scarcely requiring proof or formal statement, which almost every man, as soon as he hears them, admits as familiar to his thoughts; and his inferences are nearly as general, and if he has reasoned correctly, as certain as his premises. Those which relate to the nature and production of wealth are universally true. But these conclusions, whatever their generality and truth, do not warrant the economist in adding a single syllable of advice. That privilege belongs to the writer or statesman, who has considered all the causes which may promote or impede the general welfare. The business of the Economist is neither to recommend nor to dissuade; but to state general principles, which it is fatal to neglect, but neither advisable nor perhaps practicable to use as the sole guides in the actual conduct of affairs."

Mr Mill agrees mainly with Mr Senior in his definition of the science. It professes, he says, "to teach or investigate the nature of wealth, and the laws of its production and distribution, including directly or remotely the operation of all the causes by which the condition of mankind, or of any human society, in respect to this universal object of human desire, is made prosperous or adverse. It undertakes to set forth as much as is known of the laws and principles according to which those causes operate." "All know that it is one thing to be rich, another thing to be enlightened, brave or humane; that the questions how a nation is made wealthy, and how it is made free, virtuous, or eminent in literature, the fine arts, arms or polity, are totally distinct inquiries."

On the strength of these definitions Mr Senior vindicates writers on Political Economy from a charge sometimes brought against them. "It has often," he says, "been made a matter of grave complaint that they confine their attention to wealth, and disregard all consideration of happiness or virtue. It is to be wished that this complaint were better founded. No one blames a writer upon tactics for confining his attention to military affairs. An author who, having stated that a given conduct is productive of wealth, should, on that account alone, recommend it, or assume that it ought to be pursued, would be guilty of the absurdity of implying that happiness and the possession of wealth are identical. But his error would consist, not in confining his attention to wealth, but in confounding wealth with happiness. Supposing that error avoided, the more strictly a writer confines his attention to his own science, the more likely he is to extend its bounds."

Political Economists in general have by no means observed the rule Mr Senior lays down for their guidance, that they are not warranted in offering "a single syllable of advice." There are few articles of general commerce of which the production has been more copious and abundant. Whatever the demand for it, either on the part of statesmen or of the general public, it has usually been far exceeded by the supply. The consequence, in agreement with the maxims of their favourite science, has been a frequent depreciation in the market value of the counsels they have offered. Separate laws, and whole systems of finance and legislation, social customs, the conduct of landlords, manufacturers, merchants, and workmen, have all in turn been the constant objects of their praise and censure, from their agreement or disagreement with the supposed teaching of Political Economy, and on that ground alone. The claim that other and deeper principles ought to be kept in view, and that outward wealth is not the highest aim of personal conduct or national legislation, has been often ascribed to ignorance or perverseness, and a culpable neglect of the discoveries of this modern science. It is clearly essential to settle this internal dispute among its teachers and disciples, and decide on its true nature and limits, before we can form a correct view of its relations to Ethics, and trace the links of connection which bind the two subjects in close and inseparable union.

There are plainly, then, two opposite tendencies at work among the writers on Political Economy. One class, of whom Sismondi is one representative, endeavour to retain the true and proper sense of the phrase, as importing the right guidance and direction of the State, as a larger household. These enlarge and widen the range of the science, and make human welfare its true object. But in the same degree they render it vague as well as comprehensive, and would make it almost synonymous with the pursuit of happiness, or the highest good, personal and social, so as to include wisdom and prudence applied to the conduct of human life, in all their most varied forms. Others, like Mr Senior, confine it to the laws of material wealth alone. Viewed as a science, its proper name would then seem to be Plutonomy, or the study of the laws by which wealth is produced and distributed; and as an art,

Chrematistics, or skill in acquiring all those things which are bought and sold in the market, and which money alone can procure. The science, on this more limited view of its nature, gains in precision. But in the same proportion it becomes technical, partial, and limited in its range, and is liable to be deceptive in the highest degree, when we turn from its theories to their practical application. Two alternatives are open to its students. They may understand the science in its narrower sense, and then profess their incompetency to give practical counsel in questions of private conduct and national legislation from its teaching alone. Or they may claim the character of public instructors in all these questions, and widen the range of the science, so as to include mental as well as material riches, and all the moral elements on which human happiness depends. But the most fatal course, and one of the most frequent, is a vague, uneasy oscillation between these extremes. The science is then rigidly confined to the study of material wealth, when its theories are being formed in the closet of its teachers and disciples, and is promoted into the oracle of truth, happiness, and genuine wisdom, when it is sought to apply them for the guidance of men and nations in the actual business of life.

The true and just view of the subject lies between these opposite extremes. Political Economy, in the modern sense of the term, might be more properly called Chrematics, or Natural Polity. It consists in the research and discovery of those laws which regulate the relations between men, and all those outward objects which minister to their life, health, comfort, intellectual progress, and moral welfare. It is therefore part of a greater whole. It deals with men in their relation to the outward world. But the wealth on which it reasons, though an end with reference to the processes of production and distribution, is really only a means to a further end, the true good and welfare of men. When divorced from this higher aim, like the manna when hoarded unseasonably, it breeds worms and turns to corruption. The food becomes first husks, and then poison. The gold and silver rust and are cankered, and when the mischief proceeds further, eat the flesh as a consuming fire.

To avoid this great danger, two main aspects of the science must be clearly distinguished, and even contrasted, which too many Political Economists are perpetually confounding together. The first may be called Natural or Physical, the second Moral Chrematics. The first treats of the actual tendencies of production, commerce, and trade, when men apply themselves to the satisfaction of their own wants and desires in the use of outward objects, under the guidance of instinctive self-love or prudential selfishness alone. The second deals with a harder and higher subject, what is the true standard of right and duty in this wide field of human action, and how far the impulses of selfishness, and the calculations of mere prudence and worldly interest, ought to be restrained, guided, ennobled, and sometimes even reversed, by the higher claims of social benevolence and religious faith.

The intimate relation between Moral Truth and genuine Political Economy may be seen in three different aspects of economical science.

The first of these refers to the province of government, and the nature of those elements on which the wisdom of all varieties of human legislation must depend.

Political Economy, we are told, has for its object "not happiness, but wealth." Wise laws, however, must plainly have for their object, not wealth but happiness, or the true welfare of the nation, and the individuals of whom it is composed. The two ideas, it is admitted, are in their own nature distinct and separable. Whenever in practice they diverge and separate, the teaching of the secondary science, which deals only with outward wealth, will have to be modified, and sometimes even reversed, by the just authority of higher truth.

It follows, from the reality of the distinction, that laws may be of four kinds. They may hinder wealth, and diminish happiness, or may be promotive of wealth and of happiness also. Other, again, may encourage the growth of outward riches, and be morally injurious. Others, lastly, may restrain special modes or forms of industrial activity, by which material riches might be increased, but may still, for other and higher reasons, be conducive to the public welfare.

Free trade, when the freedom consists in abolishing laws of the first kind, is an evident gain. It releases industry from shackles by which it had been unwisely confined, which retard its progress, and limit its success, while the direct benefit is outweighed by no attendant evils. But whenever the freedom is opposed to laws of the fourth and last kind, the moral conditions are reversed. Used in this wider sense, free trade must be no gain, but a loss. It becomes a delusive watchword, under which moral opposites are confounded together. It would serve really to erect lawless anarchy of commercial selfishness into the canonized standard of private trade and public legislation among Christian men.

It may tend, perhaps, to clearness of expression to define the object of Political Economy to be the production and increase of material wealth alone. But a science so defined is limited and imperfect by its definition. In the words of ·Mr Senior, it is not warranted to offer a syllable of advice on questions of private or national duty, till it has borrowed information from other sources than its own. It must be a fatal error to promote conclusions, drawn from such limited premises, and directed to a purely secondary aim, into grand, fundamental laws of social legislation. It is an important boon that commerce should be set free from all selfish, unwise, and needless restrictions. But when, under this plea, a broad indefinite claim is set up, that it should be exempted from restraints and regulations of every kind, the doctrine becomes at once immoral and absurd. Once let the cold, hard, iron selfishness of trade, which disowns the obligations of Divine law, and even the softening instincts of human affection and tenderness, be made supreme in the intercourse of life, and an idol of the worst and meanest kind will have been set up, in the abused name of Economical Science, for universal worship.

But Political Economy is further linked with moral elements in the closest way, because it depends on personal prudence and moral conditions for the success and continuance of that industry which is the only source of private and public wealth.

And first, the moral character and state of the working classes is one of the most vital elements, on which the progress or decline of national industry must ever depend. Let us compare two communities, in one of which the workmen and labourers are sober, industrious, honest, and another in which they are drunken, idle, fraudulent and profligate, and the contrast in their social progress and comfort must be extreme. The words of ancient wisdom are true to the present hour, after all the discoveries of modern science: "The soul of the sluggard desireth, and hath nothing." "The drunkard and the glutton shall come to poverty, and drowsiness shall clothe a man with rags." The moral and spiritual state of the industrial classes has a most direct and powerful influence, for good or evil, on the whole system of trade, and on the progress or decline of national wealth and greatness. Laziness, intemperance, feverish discontent, and lawless violence, are equally fatal to the healthy development of national industry. These evils, when they abound, take off the chariot-wheels of trade, so that it drags heavily and laboriously along, and are the symptoms of some great catastrophe near at hand. And hence it has been justly observed by Dr Chalmers, in the preface to his treatise of Political Economy: "Vary its devices and expedients as it may, it can never secure its object apart from a virtuous and educated peasantry. Even for the economic well-being of a people, their moral and religious education is the first and greatest object of national policy. While this is neglected, government will only flounder from one delusive shift and expedient to another, under the double misfortune of being held responsible for the prosperity of the land, and yet of finding this to be an element hopelessly beyond its control."

But the truth, which applies to the moral condition of the working classes, belongs equally to the character of the tradesmen and farmers by whom they are employed. If these are honest, truthful, liberal and kind, the inflence in the promotion of national industry will be speedily felt. Confidence breeds confidence, kindness in the master or employer produces, as a general rule, attachment and fidelity in the workman or the servant. The wheels of commerce then move on freely, and without friction. But if the employers of labour are close-fisted, heartless, or even deceitful and dishonest, industry will be depressed and deadened into dull and lazy inaction, and the way will be prepared, when knowledge without faith has spread more widely among the people, for some sudden outburst of revolutionary violence.

When we pass upward, from the middle classes to the most wealthy, the importance of the moral element to the economic well-being of society is no less apparent. If rich men are kind, generous, sympathizing and bountiful, the unequal distribution of the gifts of Providence ceases to be a perpetual source of heart-burning and sullen discontent. Society then becomes the counterpart of that land of promise, where the hills and mountains receive abundantly the rains of heaven, but only to dispense them in refreshing streams to the well-watered valleys below. But let them once become, as a class, miserly capitalists in one generation, alternating with luxurious, profligate spendthrifts in the next, and the probable result, before long, will be an amount of sullen and morose discontent, accumulating from year to year, before which the foundations of the social system may at last give way, and wealth and industry be buried in the common ruin.

But besides the indirect dependence of national wealth on social morality, through the higher or lower standard of industry and prudence which the community and its various classes of workers have attained, there is a further connection of a still more direct and intimate nature.

Moral Science, from its very definition, extends its claim to the whole range of human action, speech and thought. It does not rest content with a partial service, the church or the closet alone. It follows men into the street, the market, the warehouse, and the field, and claims to guide and control their actions in all the various relations of commerce, industry and social life.

The words of the Apostle, "None of us liveth to himself," if viewed as the assertion of a universal fact, would be wholly untrue. But interpreted as a moral precept, they proclaim a law of universal obligation. They apply to the pursuits of trade and commerce, as well as to every other field of human activity. Purely selfish bargaining is in no case lawful for Christian men. To buy in the cheapest market, and sell in the dearest, may be a maxim of wide range in commercial transactions. But there are certain moral limitations it is bound to obey. When these are neglected or denied, it becomes immoral and mischievous. It would canonize the hideous principle that, in the affairs of trade, they were at liberty to put to sleep every higher instinct of humanity, and to be guided by the promptings of pure and absolute selfishness alone. The maxim, "caveat emptor", may embody the results of a large experience. But its full and perfect development must require a world of devils. Its traders would be men whom the fear of discovery alone deters from the grossest frauds, and who have the balances of deceit perpetually in their hands.

The reasonings of political economists often rest on a secret assumption, that men, in all relations of worldly business, are actually guided, and may lawfully be guided, by self-interest alone. Now, even as a statement of facts, the simplification is excessive and untrue. But when accepted as the ideal basis of a science, it becomes an outrage on the simplest lessons of true morality. On such a foundation of universal selfishness a perfect industrial state can never possibly be reared. For the purely selfish in aim can never be far-sighted. The nearer present, the seeming immediate gain, will eclipse from their view the loss which is sure to attend, soon or late, on dishonesty, violence, and oppression. But shortsighted selfishness is fatal and ruinous in trade and commerce, no less than in the higher fields of politics and religious faith. Its grapes are grapes of gall, and their clusters are bitter. For credit, that is, faith between man and man, is the life and soul of commerce. But when selfishness reigns, and sits enthroned on the altar of perverted science, the atmosphere of the whole world of commerce becomes mephitic and stifling. Frauds will multiply and increase without limit, and credit will grow sickly, and soon expire.

But we have not yet exhausted the lines of thought, which prove the close and vital union between Morals and Political Economy. It may be traced further and deeper still. The truths to which we are led by further search will prove the imperfect and provisional character of many conclusions, which modern writers on Political Economy are accustomed to invest with a character to which they have no just claim, not as the results of an imperfect hypothesis and a rude approximation, but as absolute and undoubted truths of a perfect science.

I have reasoned hitherto on the assumption that wealth can be defined, wholly apart from every moral element, and that moral truths are needful only for its right use and application. But the admission, made for the sake of argument, has no real foundation in truth. True Wealth, by the very origin of the name, is not wholly distinct from "welfare," but involves the same idea, embodied and transformed. It consists really in the outward means of welfare, when possessed and employed by those who have the wisdom to use them aright. The word "goods" has a like derivation, and points to the same lesson. Things are not goods by virtue of a process of manufacture alone, but because they are applied, or may be applied, to promote the real good of those who own them. The right application of the products of industry is the secret condition, too often overlooked, on which their character as wealth really depends.

The idea of value lies at the foundation of Political Economy, just as the conception of space is the basis of geometry. And yet the definition of value has been one chief perplexity of writers on the science. One eminent writer confounds it with cost, and has thus involved his whole treatise in hopeless perplexity and confusion. Many identify it with market price. Another, in a separate treatise on the subject, denies it any fixed meaning, and looks upon it as a mere term of relation, in which one article rises and another falls, but no absolute measure is possible. Mr. Mill boldly affirms that nothing remains for any future writer to clear up, and that "the theory of the subject is complete." It may be affirmed, with more truth, that its first elements are hardly laid, because the main principle on which a true understanding of it depends is overlooked or forgotten.

The source of these perplexities is not hard to discover. It arises from a persistent effort to separate the *body* of wealth, its outward and familiar form, price in the market, or a sum of money, by which anything may be purchased, from its *soul*, without which it ceases to be wealth, and becomes a worthless corpse, a glut of unused, or an excess of abused commodities. That *soul* of wealth, without which things are things only, and not goods in the proper sense, is their real application to some end, which is beneficial to the owners. When this condition is not fulfilled, wealth ceases to be wealth, having nothing to do with human welfare, but becomes either waste or poison.

The whole of economical science may be said to depend on two or three fundamental equations, which are the moving forces in the whole system of trade. Cost, increased by a first profit, is the market value. Market value, increased by a second or third profit, is the worth, or value in use. But cost may be either real or imaginary. There is the same contrast of true and imaginary worth. The mechanism of trade depends on the cost and worth, as defined by the mere fancy, often the erring fancy, of the producer and consumer. But the benefits of trade depend on a comparison of the real cost and the real worth, and of these alone. The great and all-important contrast between a cost and a worth which are real, and one which rests on erring and mistaken impressions only, is the principle on the recognition of which it depends whether Political Economy shall be a moral or an immoral science.

Mr Mill, for instance, lays down the principle that "Political Economy has nothing to do with the comparative estimate of uses in the judgment of a philosopher or moralist. The use of a thing, in Political Economy, means its capacity to satisfy a desire or serve a purpose." Here the fundamental defect comes clearly into view. The object of Political Economy, as thus defined, is to multiply the production of things desired, however vicious, hurtful, and even ruinous, the indulgence of those desires may be. If men are swine, its object is to provide more husks and refuse for the troughs in which they feed. If they degenerate into devils in malice and hatred, it is to multiply the desired engines of mutual destruction.

But a science thus defined, and violently severed from all consideration of right and wrong, of the true welfare of man, or hurtful delusions that work his greatest harm, can be no science of human action. It may amass materials, from painful and humbling experience, to teach what fools, misers, and profligates are likely to do. But it renounces the attempt at such an estimate of the wise and just objects of human pursuit, and of the right use of the ample materials for happiness the God of nature has spread around us, as can alone form the basis of a genuine science, and fit mankind to exercise wisely their sacred trust of ownership over this lower world.

Etymology is often a guide to truth, when it is overlooked by those who are busied in framing some artificial theory. The term now in question is an instance of this kind. Value has a double relation to weight and health. It denotes, first, what anything avails in the sober estimate of sound reason, when it weighs men, actions, and things in an even scale, and rates the last according to their real worth, or power to promote man's solid and lasting welfare. And it implies, further, a quality in outward objects, akin to health in the human body, which fits them for the discharge of some useful and healthy function in the complex economy of human life. And no Political Economy can advance human happiness, or escape from the risk of nursing mischievous delusions, which does not accept the teaching of a nobler science than its own, and draw a wide contrast between substance, which has weight, and shadow mistaken for substance; between objects which satisfy the desires of health, and those which minister chiefly to the restless cravings of indolence, miserly greed, destructive profligacy, and all the varieties of moral disease.

The vast importance of the truths here advanced, and neglected or disowned by many writers on economical science, has received, and is receiving at this hour, a striking illustration from the course of modern history.

Since the days of Adam Smith, the founder or reviver of Political Economy in its modern form, there have been two successive stages in the history of its progress and effects. The first stage seems nearly completed. The second, its converse, has only just begun.

In the first stage, which owes its development to many distinguished writers, there has been much acute and sagacious, and also not a little fallacious reasoning, on the tendencies of trade, and the laws of the production and distribution of wealth, so far as determined by motives of pure commercial self-interest alone. The science, on this view, is one neither of mere practical observation, and inferences, *à posteriori*, from the observed facts of trade and agriculture, nor yet of moral duties, and what course men, in the commerce of life, *ought* to pursue. It is a hypothetical science of the tendencies of trade, assuming self-interest, which has doubtless a very wide range, to be the only motive in ceaseless operation. Such a study of the tendencies of commercial self-interest, whether sinking into hard, cold, shortsighted selfishness, or rising into that moral prudence and discretion which is one of the main Christian virtues, may be highly useful, if honestly pursued, and kept within its own proper bounds. But in this case, speaking generally, a double mistake was made. To render the inquiry definite, like those of geometry, moral elements were thrust out of sight, and the question whether things had real value, or were only conceived to be valuable through error and ignorance, was ruled to be entirely beyond the range of the new science,

And still it was forgotten that this exclusion thrust it down to a position almost servile. It became a theory of the tendencies of commercial selfishness alone, and was guilty of plain usurpation, when it affected to lay down imperative rules for private conduct or national legislation. And next, the inquiry into tendencies themselves was warped and distorted by this confusion of thought. There was a constant disposition to idealize that self-interest, which was accepted for the supreme law of trade, to purge it from the scum of those grosser forms of selfishness, with which it is always largely combined in actual experience; and to leave out of sight those two mighty sources of unknown disturbance, the selfishness of the rich and covetous, when capital accumulates, and the rival selfishness of discontented multitudes, when the contrasts of society widen into a chasm, bridged over by no instincts of humanity, and by no common hope of a nobler life to come.

That selfish school of Political Economy, which dates from Adam Smith, and is really a Hypothetical Science of the tendencies and laws of commercial and productive self-interest, seemed to prosper greatly in its first stage. Trade was gradually set free from many hurtful restraints, some of them possibly useful in former days, but grown antiquated and worthless, and therefore mischievous, and others, even in their origin, due to some form of blind selfishness or jealousy alone. The new discoveries of the steam engine, the steamboat and the railroad, and the opening out of the vast hidden treasures of force in our coal measures, concurred in the mighty change. There was an immense increase of productive power. Labour was partly displaced, but still more stimulated and quickened, by a vast variety of mechanical inventions and discoveries. It was an era of growing capital. Our national wealth and resources were greatly increased. But side by side with this gain was the dark shadow of a great and growing evil. Along with the growth of wealth there was a growing inequality in its distribution. Acquired wealth, under an accepted rule and law of pure selfishness, and when capital is worshipped as the chief good of a nation, has increasing power to beat down the price of labour, and secure the main share of its fruits; until a rival power, still more dangerous, the brute force of myriads and millions, banded by common discontent, and ready to look upon all the wealthier classes as cruel and worthless oppressors, arises to struggle for the mastery, and to transfer the sceptre of trade to other hands.

That second stage in the history and development of a selfish commercial philosophy has now set in. It would have visited us long ago, if the defects of the new science had not been counteracted by a large amount of Christian charity and kindness, softening the hard rigour of the economists by lessons of a deeper and holier kind. The factory laws, and other kindred efforts, proved that spurious lessons of a maimed, imperfect science had not wholly paralyzed the national conscience, and destroyed the instincts of humanity. But, in spite of many things which have delayed its advent, the later stage has arrived, and a reaction, long seen to be inevitable by thoughtful minds, has begun. A new name, the proletariat, has been invented for the working classes, and represents the widespread impression among them, that they have been used by the selfishness of wealth as mere machines for the production of riches, and hardly been regarded as fellow-men and fellow-Christians. By the spread of a cheap literature, and the means of rapid locomotion, their opportunities for combining together have been greatly increased. The separation of remote towns and provinces has been replaced by one of another kind, the sharp contrast of classes, differing in social level. Trades' unions, national and international societies of working men, have been formed on a wide scale. Strikes, concerted by their leaders, and carried out with a military discipline, and under the terror of punishment little short of death, have become of yearly and almost hourly occurrence. Communism in theory, or a total denial of the right of property, has spread widely, and fascinates tens of thousands with the hope of an entire social revolution. The selfishness of capital, in the ascendant for two generations, and almost consecrated into the chief of virtues by modern system-builders, is now encountered and confronted by a force still more terrible, the organised discontent and self-will of the most numerous classes, on whose willing labour the whole system of trade and commerce hourly depends. From the operation of these new forces our great metropolis has had a narrow escape from the double danger of midnight darkness and winter starvation. In one word, Political Economy, in the hands of many of its chief writers, has committed the fatal mistake of affecting to be independent of all moral truths, and still to constitute a double guide for personal conduct and for national legislation. "The greatest happiness of the greatest number" has been its professed aim, and "the greatest discontent of the greatest number" would almost seem, from the present signs of the times, to have been the result actually achieved.

What, then, is the remedy for this social crisis, which threatens to replace one great evil by another still more perilous, and perplexes both monarchs and republics, at the present hour, with the fear of disastrous change? The efforts of many hearts and minds, the unfolding of many high truths, the culture of many Christian virtues, are all required for this hard task. But one main duty is plain. We must "moralize" Economical Science. We must relegate the science which has usually borne that name to its true, but subordinate place, as a science of tendencies, not of duties; an hypothetical, not an actual science, because it deals with the working of one class of motives only, and those neither the highest nor the best, those of self-interest and money-making alone. Side by side with Natural Chrematics, which has usurped the higher title of Political Economy, we must create and develop a higher science of Moral Chrematics, or the laws of social duty, by which man is bound both to God and his fellowmen in the right use, social adaptation, and religious consecration, of all outward and visible things. It is not universal free trade, least of all its freedom from moral restraints, but universal uprightness, integrity, and brotherly kindness, in trade, labour, commerce, and all social relations, which is the true and effectual remedy for these threatening evils. The whole system of trade, the manifold treasures which human skill and industry derive from their parent earth, when once they break loose from the great law of love, may soon lose themselves in outer darkness. But when they learn to submit reverently to this higher law, they revolve harmoniously in their own orbit, and are bathed in a light and beauty which shines on them from above. And the first lessons and definitions of a Political Economy, thus reclaimed to its true dignity, and placed on firm and moral foundations, may be found in four sentences of the Divine law, which form a climax of heavenly wisdom: "Six days shalt thou labour, and do all thy work." "A man's life consisteth not in the abundance of the things he possesseth." "If ye have not been faithful in the unrighteous mammon, who will commit to your trust the true riches?" "Charge them that are rich in this world, that they be not highminded, nor trust in the uncertainty of riches, but in the living God, who giveth richly all things to enjoy: that they be rich in good works, ready to give, glad to distribute, laying up in store a good foundation for time to come, that they may attain eternal life."

LECTURE VIII.

MORALS AND POLITICAL SCIENCE.

THE intimate connection between Ethical and Political Science has been generally recognized from the time of Aristotle down to the present day. But it is not very easy to attain a clear view of the exact nature of the fundamental relation which exists between them. The path of truth and wisdom seems to lie between opposite bye-ways of shortsighted, prudential selfishness, and of mere utopian theories, and it is only too easy to wander from it on either side.

In every field of human thought, which aims at the guidance of practice, truths and facts, science and experience, *à priori* and *à posteriori* elements, are two factors which need to be combined, and seen in their mutual harmony. This is eminently true in Political Science. It cannot receive a safe and full development, unless we recognize the authority of great ethical truths, or at least the guiding and controlling power of deep moral instincts, side by side with the cultivation of practical statesmanship, and of those lessons of a wise expediency, by which the laws and customs of every nation are adjusted to the varying wants and altered circumstances of society from age to age.

There are thus two opposite dangers to be avoided, in tracing out, practically, the relation between public legislation and the principles of moral science. Nations and their rulers are liable, on the one side, to be wrecked on the Scylla of some impracticable theory, mistaken for a moral law of binding and eternal obligation. Or else, in shunning this danger, they may be sucked into the Charybdis of an expediency, devoid of all social principle or religious faith, in which nations float blindly at the mercy of the winds and the waves. Principles, laid down without warrant as fundamental laws of political duty, may encroach on the just and proper range of social expediency. The firmness and stedfastness of moral truth, in its own nature, is then transferred to social customs, national laws, and political constitutions, and a diseased fancy clothes them with that immutability, which is really a Divine prerogative. Out of this error there naturally grows up a superstitious and blind conservatism, which forgets the true saying of Bacon, that Time is the great innovator, worships the dead forms of the past, and overlooks the fact that national life implies growth and change, and that altered conditions impose new duties on each successive age. The other evil, often caused by recoil from the first, is perhaps wider and deeper still. It is when mere expediency encroaches on higher ground, where fixed principles and laws of social and religious duty should reign supreme. All national sense of moral right and wrong, the sacredness of treaties and covenants, the axioms of public duty and religious faith, are then flung aside, and the place left vacant by their removal is filled up by a blind subservience to the momentary impulses of popular self-will. The true standard of excellence is forsaken and reversed, which was laid down by a heathen poet for the guidance of the upright statesman :—

<blockquote>
Virtus, repulsae nescia sordidae, in-
contaminatis fulget honoribus;
Nec sumit aut ponit secures
Arbitrio popularis aurae.
</blockquote>

Such a diseased liberalism, which neglects every higher and deeper law of national duty, and flatters and obeys the momentary impulses of public opinion, that fluctuate from hour to hour, is inconsistent with all political stability. It thrives in an atmosphere of excessive and ceaseless change, and builds on the loose sand, which the last tide-wave of the ocean may have heaped on the shore. To shun these opposite errors on the right hand and the left is the hardest and highest triumph of true political wisdom.

Moral Science, when rightly pursued and clearly attained, reveals those fixed laws of duty towards God and men, which are binding at all times and in all places, not only on individuals, but on communities and nations. They are the first elements and dominant truths in a vast scheme of moral government, which is in ceaseless operation. They are the stars of the moral firmament. They live, in the words of Sophocles, "not only yesterday and to-day, but evermore, and no one knows any hour when they first were born." Whenever these are transgressed, however strong may be the temptation, and however plausible the excuses by which men or nations try to conceal their shame, the penalty is sure, and the stars, in their eternal courses, fight against the transgressors.

But laws and customs, however firm their moral basis, need to be adapted to the wants of successive ages, and of the various tribes and races of the human family. The general duty requires to be embodied in special details, for practical use in the affairs of life. And these details of social law and usage are mutable in their own nature. They may and ought to be "changed according to the diversities of countries, times and manners." And true wisdom, both in laws civil and ecclesiastical, consists mainly in the avoidance of "two extremes, too much stiffness in refusing, and too much easiness in admitting" the variations by which they are fully adapted to the actual state and wants of human society.

Now there is a moral superstition, often very mischievous in its results, which confounds the secondary and derivative laws with those primary lessons and principles of good and evil, of right and wrong, on which they are secretly founded. The error, in political thought, bears a close resemblance to the doctrine of transubstantiation in theology. The outward sign and visible clothing of moral truth is confused with the truth itself, till the form is blindly worshipped, and the substance obscured and lost. The natural result is some form of political idolatry. But the nature of the idol may vary from time to time. Sometimes it is an antiquated and worn-out usage, sometimes a particular flag, or family, or some worthless pretender to a throne, long ago forfeited by human guilt and folly, when the just sentence has been sealed by Divine Providence through successive generations.

To see clearly, then, the true relation between Ethics and Politics, the first essential is a firm hold on the contrast between Pure and Applied Morality. The great truths, which appeal directly to the heart and spirit and conscience of man, have to be distinguished from their application, in detail, to all the varying forms and conditions of social life.

The laws of truth, love, righteousness, and grace are

of absolute and universal authority. They belong alike to all moral agents—to men, to angels, and to unknown races, distinct from both, if such there be; and to men in every age of the world, of every race and tribe, and under all the varieties of political constitution or social change. But their application, in social jurisprudence and outward law, admits of large varieties. It depends partly on experience of what men have been and are, and of their actual state in moral capability, progress, and attainment, and not simply on what they ought to be. Laws, that would be admirably suited to a state of high moral feeling and advanced intelligence, may be wholly unsuitable among a race of savages, and in a state of social barbarism and abounding passion. They may then even become mischievous, and only aggravate the evils they attempt to restrain or cure.

>Quid vanæ sine moribus
>Leges proficiunt?

This first contrast lies thus at the basis of all wise and wholesome legislation. Pure Morals are fixed, absolute, invariable. There can be no curvature whatever in a straight line. But Applied Morals, in their own nature, and by virtue of their definition, are largely variable. For they include all those social conditions, as elements in their wise adjustment, which themselves vary widely, from country to country, and from age to age.

But there is a further contrast, more liable to be overlooked, which is also highly important, and even vital to a just and comprehensive view of the whole subject. Positive Law or Applied Morals has two different aspects, one human, and the other divine. And these are distinct in their aim, and involve important differences in their practical development.

The object of outward laws, in their human aspect, is merely the repression of vice and crime, so far as they tend to injure or destroy the life, comfort, and social welfare of men. Its leading and defining character, then, is to be restrictive of evil.

The Divine aspect of Law goes further. It includes, indeed, this first and more immediate object, the repression of open crime. But it has a wider and deeper range, and includes, as one main end, the discovery of evil, with a view to its more perfect abatement and cure. It enters "that the offence may abound." It is not simply a power and force to restrain evil. It is detective also. It does not confine its aim to the security of life and property for the moment, but reaches after the higher object of moral renovation.

Human laws themselves, when guided by true wisdom, as soon as the first end is secured, should pass on to the higher. The first endeavour must be the repression of outward vice and crime, in those forms which strike directly at the peace and comfort, and even the existence of society. The second and higher is the awakening of conscience from its slumber, and the creation and culture of a higher moral standard. In this way laws themselves may become one powerful means of moral education.

Positive Laws, when divinely revealed, may deal more largely with the higher and nobler object. Their seeming failure, when viewed only as expedients to hinder the growth of open crime, may be really the fulfilment of a higher object of God's moral government. The discoveries of human evil, which through successive ages they have not overcome, but rather have placed in a clear and full relief on the page of history, may prove the best preparation for more effectual displays of the Divine goodness, and of redeeming love. But such laws, when only of human origin, must be content with a more limited aim. They cannot, indeed, without serious evil, entirely set aside the higher object. But it needs to be cautiously and temperately pursued. It is a half truth, though it is only a half truth, that you cannot make men moral by acts of parliament. Human laws should be in advance, but can only with safety be a little in advance, of public morality. Else, in attempting too much, there will be serious risk, almost the certainty, of entire failure. Instead of producing habits of virtue, they will then breed a plentiful crop of mischievous and odious hypocrisy.

Legislation, then, to be practically perfect, requires a real accommodation of human laws to the actual moral attainments of society, while still avoiding with care any sanction of immorality. Like the law of Moses, but on a still more extensive scale, it must suffer some things, without inflicting penalties, because of moral blindness or hardness among the multitudes for whose good it provides. And hence the true principle by which it should be guided is neither moral despotism, nor moral indifferentism, extremes which tend to generate each other; but Toleration, or the public recognition of a contrast between what the law condemns and punishes, and what the lawgiver fully approves. It is a watchword, which bigotry may dislike, and charge with moral weakness, and which may be resented as an insult by the pride of lawless self-will; but no other principle, in a world like ours, can satisfy the imperative claims of high-toned morality and true political wisdom.

A third main element, essential to a just view of the intimate connection between Morals and Political Science, is a clear apprehension of those Formative Principles, on which the right and wise establishment of every system of positive law must depend.

Political Law deals with men under three distinct, but closely related characters, as Men, as Children, and as Brethren. Hence in all social law there will be found to co-exist three main and vital elements,—Humanity, Piety, and Brotherhood. Or, in other words, to bring the truth into closer relation with a favourite modern watchword, with which it agrees in part, and partly disagrees, its three principles are Liberty, Inequality, and Fraternity. Neglect of any one of these is mischievous. Its deliberate and entire reversal, whenever seriously attempted in the social constitution, must be fatal and ruinous.

The first of these essential principles of wise legislation is that of Humanity or Liberty. Men, in order to be the fit subjects of social law, must be free moral agents. The freedom thus affirmed is no contradiction of that unhappy bondage to habits of sin, which Scripture and experience alike reveal as a mournful and widespread disease. On the contrary, it is the condition without which moral health and moral sickness are both impossible. Bondage of such a kind is the disease of agents, morally free, and is possible to these alone.

Laws, then, being designed for men as free moral agents, must deal with them according to their true character. If they are treated as mere things, chattels of a master, or mere tools for the manufacture of goods, whether this be done in the name of political absolutism, commercial selfishness, or a fatalistic and material philosophy, morality is quenched, the hearth of human affection is left cold and desolate, and social life must expire and die away.

The evil, to which this great principle is opposed, may

assume at least three different forms. In philosophy it reveals itself in that cold and barren fatalism, which would resolve the heart, mind, and conscience of man into a transcendental product of solar force; and pretends to believe that our feelings of right and wrong are bound hand and foot in the toils of some vast dynamical theorem, which awaits the birth of a competent intelligence in one of its own countless variables, for its own complete integration. In politics, the same evil appears in the two different forms of mere absolutism, when all rights of subjects are displaced by the caprice of a supreme despot; or of caste absolutism, when a claim to lifelong service is perverted into a denial of all the rights of humanity, and instead of servants, more honourable than mere hirelings, who honour their masters, and receive from them what is just and equal, we have selfish oppression on one side, and on the other a race of degraded helots and slaves. The same evil reappears even under the outward show of extreme liberty, when men and children are treated by their fellows as mere tools and instruments of production, and all their higher characters are lost sight of and forgotten. A third form of the same evil is found in hasty over-legislation, when human lawgivers attempt a task beyond their power, and in the pursuit of some political object, or in the mistaken assertion of their own authority, bind down the social life of the nation with burdens hard to be borne, which experience proves to be intolerable, and which, when long continued, lead to discontent and social revolution.

This first principle, then, of Humanity or Liberty, prescribes a threefold law for the guidance of all social thought and action, and of all wise legislation. First, in doctrine, the clear assertion and firm maintenance of human responsibility, in contrast to the immoral voices of a material philosophy, and even to those subtle physiological excuses for crime, which forget that all wrong doing, when traced deep enough, must be real insanity and madness in the sight of perfect wisdom. Next, in the national constitution, the same principle points to power limited by law, or the settled order of constitutional government, in contrast to the despotism either of lawless multitudes, or of some solitary tyrant. And lastly, in practical politics, the same truth points to the duty of large toleration and forbearance. It will lead statesmen to remember the limitations of their own power, and to avoid the serious risk of over-legislation. They will feel it not only lawful and expedient, but even one primary obligation of their high office, to leave wide scope for human freedom, without which life becomes mechanism, and society degenerates into a herd of slaves.

The second formative principle, on which the moral character of human legislation depends, and by which it ought to be controlled, is Piety, in the primitive and classic meaning of the word, which includes reverence to human superiors, and especially to parents, as well as towards God and the unseen world. It is the principle, in other words, of Inequality or Subordination.

Men, by the very constitution of their twofold nature, are born unequal. The opposite statement, however widely current in modern times, is either a strange misnomer, or a prodigious fable. Each child of man is first a child, and only much later becomes a parent. A contrast, wide and deep, between experience and inexperience, authority and subjection, manly strength and infant weakness, is inwrought into the constitution of human nature, and reveals itself anew in every successive generation. There is no strict equality even among the hosts of heaven. There are angels that excel in strength, the rulers of celestial hierarchies, thrones, dominions, principalities and powers, dimly and mysteriously revealed. Far less can such equality be found among men. It has no place, still lower in the scale of being, among the animal creation and the vegetable world. We must stoop lower still, to the first elements out of which lifeless objects are composed, and there alone, if anywhere in creation, equality must be found. The atoms of chemistry are unequal, from hydrogen to gold. Comets and asteroids, satellites, planets, and suns, the hyssop and the cedar-tree, the insect and the elephant, reveal the law of inequality in the natural world. And it extends still further, and mounts higher, in the contrast between the newborn infant, the strength of manhood, and the wisdom of age; between the drunkard and the savage, and saints, and prophets and apostles, the lights of the Church in every age. It is thus one all-pervading law of the natural, the intellectual, and the moral universe.

> Order is heaven's first law, and, this confessed,
> Some are, some must be, greater than the rest.

Strict and absolute equality would unmake the universe, and resolve all its wonderful beauty and endless variety into a chaos of warring atoms and a sea of nebulous mist once more.

This law of order, which underlies the whole world of political life, and binds it to the great doctrines of ethical science, assumes the double form of social and religious piety. It involves, in its manifold and weighty lessons, the subjection and obedience of children to their parents, the reverence of the young to their elders, the subordination, in loving union, of woman to man, and the dominion exercised over the floating, variable, and inconstant wishes of a people by fixed and abiding laws of national life. Its higher form is seen in the dependence of the whole community on the will and word of the Supreme Creator. And the two great evils by which the principle is disowned, and political science is disjoined and severed from moral truth, are social insubordination and national irreligion. The first of these evils is sometimes a penalty, which follows upon the second. When religious faith declines, and its sanctions lose their power, there follows, naturally and inevitably, a relaxation of those bonds which form the cohesion of society, and bind it together. The warning description of the prophet is then fulfilled. The child behaves itself proudly against the ancient, and the base against the honourable.

The third moral principle, which is inwrought into the whole texture of human society, and prescribes a law to the form and course of wise national legislation, is that of Brotherhood or Fraternity.

Men are born children, but they grow up to be brethren. The brotherhood of a common nature, of common wants, rights, and duties, of capacities for moral progress, and dim yearnings after some higher happiness still unattained, exists side by side with all the inequalities of age, sex, talent, wealth, social position, intellectual gifts, and moral attainments.

This grand truth of man's universal brotherhood limits and guards the other maxim of inequality and subordination, and hinders it from lapsing into a serious error. The two principles, like polar forces in the natural world, balance, and in balancing strengthen each other. The peasant, the workman, the pauper, the newborn infant, is a brother or sister to the wisest philosopher, the most gifted poet, the most absolute sovereign; and to those who are higher than mere philosophers, poets, and kings, to those prophets and apostles, who in moral dignity and honour are highest and noblest among men, who stand in the presence of the Lord of the earth, and convey His messages with authority to their fellow-men. And this brotherhood has its rights, which cannot be violated without guilt and danger, and must form one main guiding law in the whole course of wise legislation.

The evils in political life, which contradict this great law of social brotherhood, are many and various. First in order is the selfish pride of rulers and governors, when they look upon their office as a means for personal aggrandisement, whether in the grosser form of wealth and opulence, or the more subtle shape of ambition and historic fame; instead of seeing in it a sacred trust, to be used for the benefit of their fellow-men. This evil reveals itself outwardly in the form of caste legislation, or the tendency to confine the main benefits and privileges of the social system to a favoured minority alone. The second, near akin to the first, consists in the pride and exclusiveness of social distinctions. The lesson of order and gradation is thus exaggerated, till it hides from view the common dignity of our human nature, and the further claim of affinity and brotherhood. A third evil is the degradation of woman. The recoil is easy, from the lessons or customs of a high-flown chivalry, which treats woman, in terms, with a fond and blind idolatry, and sets aside the truth of her subordination, so clearly revealed, to a refined, but debasing sensuality; and then, as the moral darkness increases, to a coarse, ungenerous and heartless abuse of men's superior power, till the object of idolatrous enthusiasm to the knights of romance, or a sentimental philosophy, is turned once more, as in savage tribes, into a victim and a slave. A fourth evil, akin to the last, is the oppression of childhood. Some painful forms of this evil were too frequent among ourselves, till they were checked by wise and Christian legislation. Others, it is to be feared, survive still to the present day. When idle and drunken parents indulge their own vices through the hard earnings of their young and almost infant children, a blight must have passed over some of the sweetest and holiest instincts of our nature. The aggravated guilt is too likely to be followed, in later years, by the opposite and not less hateful evil of childish lawlessness and disobedience, till the springs of family life and love are poisoned and destroyed.

The last and crowning denial of human brotherhood is that of direct slavery. Only it should be well observed that the definition of slavery, which brings it under this just sentence, is not a legal obligation, on the part of one man to another, of lifelong service. It is the treatment of man by his fellowman as a mere thing, and not a person, a tool for producing wealth, and not as a moral agent, sharing the same nature with himself, and forming a true part of the wide and sacred brotherhood of mankind.

These great principles, on which the connection between Ethics proper and Political Science seems really to depend, involve a large variety of weighty practical lessons. We may apply them, in succession, to the duties of subjects, of magistrates, and legislators; and, rising still higher, of those who attempt to frame or remould that constitution in any state or kingdom, which may decide and determine the whole course of future legislation for ages to come.

The first duty, then, of subjects, which flows directly

from the first of these truths, is to be freemen indeed. They are to be subject, not to lawless caprice, under whatever titles of honour or sacred names it may be disguised, but to law and just authority alone. This is their birthright, and they cannot without guilt and shame barter it away. This truth is the living root of all constitutional government, in contrast to political or ecclesiastical slavery, the mere despotism of Pope or Emperor, of tyrant or priest. No man, whatever his rank, spiritual or political, has the right to claim absolute, unlimited obedience from his fellow-men. Those religious orders, which lay down such a rule for their members, may gain an effective organization, so powerful as to make them dangerous to their enemies. But they purchase it by sacrificing Christianity itself, and stifling the voice of natural conscience. He who consents to be the slave of a superior or a priest ceases thereby to be a freeman of Christ. And the evil is nearly the same in abject submission to political tyranny. By such a blind subserviency the cause of order gains nothing. It rather incurs a serious loss. Doctrines like those of the *Leviathan*, with the corresponding practice of political servility, form the natural prelude to those fierce and bloody revolutions, in which all the foundations of social order are overthrown and swept away.

But if the first duty of Christian subjects is to be freemen, who bear in mind the dignity of human nature, not destroyed, though obscured, by human sin and guilt, and which constitutes the very basis of human society— their next is to be men of order, or law-honouring, and law-abiding freemen. They are bound to render to lawful government, and all the authorities placed over them, no reluctant and enforced submission, but a willing and cheerful obedience. The outward acts of submission are due to Cæsar, and may be enforced by Cæsar's law. But the free and cheerful spirit in which their obedience should be rendered lies beyond the range of Cæsar's power to enforce and secure. It is a debt which is due to that higher Power, by whom kings reign, and princes decree justice. Only they who obey human laws in this temper fulfil the nobler and more sacred duty which they owe to their Maker, and "render unto God the things which are God's."

But these two principles of order and liberty, contrasted yet harmonious, do not exhaust the moral elements which enter into the wide range of the political duty of every citizen in his public life. From the truth of human fraternity or brotherhood a third main duty must follow. They ought to be men of kindness. Their moral obligation, as members of society, cannot be satisfied by obedience to the mere letter of human laws, however undeviating and complete that obedience may be. There are other and deeper laws of brotherhood, which find no place in human statute-books, and would flee away, like the delicate colours of the rainbow, if hardheaded lawyers strove to draught them in legal phraseology, and embody them in some supplement to a human code. But though unwritten, they are of real and binding obligation. In the Divine law two or three cases are written down as specimens of the rest. "Thou shalt not see thy brother's ox or his sheep go astray, and hide thyself from them; thou shalt in any case bring them again to thy brother. Thou shalt not see thy brother's ass or his ox fall down by the way, and hide thyself from them; thou shalt surely help him to lift them up again." But this law of brotherly kindness is too wide in its range, and too pervasive in its fragrance, to be fully expressed in any code of written obligations. Whenever such attempts are made, that fragrance exhales and begins to pass away. It finds its development in a thousand instincts of high-souled chivalry, and in countless amenities and courtesies of social life. And whenever it is ennobled and purified by the higher lessons of Christian faith, it raises human intercourse from the mere mechanism of law, or the drudgery of mere workers in a vast hive of industry, into a lively pledge and earnest of all the sacred fellowship which Christianity assures to the faithful in a happier and holier life to come.

These same principles, which involve the chief laws of moral duty binding on subjects in their political and social life, apply no less to magistrates and lawgivers themselves. They ought so to legislate, that these truths may be embodied in the whole course and tenour of their legislation. And this will lead them to avoid three main evils, which it is the task of true political wisdom to resist and put away.

The first of these is over-legislation, with its natural and inevitable tendency to legal tyranny. To meddle with social intercourse at every point is not the task of wise human legislation. Such a prerogative belongs to Divine enactments and the deeper laws of moral duty alone. But in the laws of men a wide scope must and ought to be given to human freedom. The true limits may be hard to define. Mr. Mill's treatise on *Liberty*, while it contains much that is true and important, lays down the principle too broadly on the other side. It assumes that personal convictions, if sincere, absolve from all right of social restraint, not only on the opinions themselves, but on the social acts to which they lead.

And it justifies this view by an appeal to many instances of social persecution, which, even when most mischievous, have been honest and sincere. But it is plain that the rights of individuals on one side, and of society on the other, depend on the real truth, and not on false opinions which either may entertain. The problem is too hard and vast to be solved by general aphorisms, investing either individual conscience, or human society as a whole, with a factitious supremacy, to which truth and liberty are sacrificed on either side. So much, however, seems to be practically clear that, in an age like our own, over-legislation must be more than usually dangerous, and that many evils should be wisely tolerated, so far as direct legislation is concerned, lest attempted remedies should be worse than the disease. But whenever special forms of selfishness or vice are proved to work extensive social mischief, and the public mind is prepared for some decisive action, it is dangerous, on the other side, in obedience to any abstract theory of personal rights, to neglect the just use of authority for the restraint of evil. It is a most unwise antithesis when it has been said that to "England sober" we should prefer "England free," as if the spread of national drunkenness were not, in its own nature, adverse and fatal to the only freedom which deserves the name.

The second evil, against which these principles ought to be a moral safeguard, is that of blind, selfish, and democratic levelling.

Human equality, unless the phrase be so limited as to become wholly unmeaning, is a great and evident falsehood. Legislation, whenever it is based on such a principle, must be a building of sand. Whenever it deals with men as mere units, and leaves out of sight the deep contrasts and immense varieties of human life, it tends rapidly to political chaos and confusion. For our very conception of chaos is that of a state in which countless atoms

> Swarm populous, unnumbered as the sands
> Of Barca or Cyrene's desert shore,

but fight only under their party colours, and have no law of mutual cohesion, no gradation of life and dignity, but decide their quarrels, ever renewed, by shifting numbers alone. Wherever such a policy is pursued, the instinct of order, violently repressed during the chase after an impossible ideal, is sure to break out in some new and unexpected form. Plebiscites of mere floating political atoms, in which every law of social and moral order is cast aside, and number alone is supreme, swiftly breed despotic rule as the natural and almost necessary antidote for a worse evil. Plato's genealogy of the forms of government is then fulfilled. The nation, which drinks deep of licence under the name of liberty, falls speedily under the yoke of some skilful and prosperous tyrant.

A third and last evil on the part of rulers, from which these great moral principles of humanity, piety, and fraternity should be a constant safeguard, is the more subtle, but hardly less dangerous fault, of class legislation. This fights plainly against the great law of human brotherhood. When the interest and welfare of large classes of men are systematically set aside, or even carelessly overlooked and forgotten, because legislators are selfish, and abuse the power vested in their hands, there is sure, soon or late, to be a dangerous recoil. Their sin is great in abusing a sacred trust, and it will be sure to find them out. They are bound, as legislators, to consult impartially for the general good, and the omission of any class from their study for the common welfare is a serious wrong. It is true that they are bound to consult for all, not as mere abstract units, or apart from all the distinctions of sex, age, knowledge, character, on which the constitution of society depends; but for all alike in their true place, and according to that weight which they may justly claim in the political organism, and in God's wide plan of universal providence. All men are not equal, far from it. But all are equally brethren. The rights of brotherhood belong to all, from the least to the greatest, and cannot in any one case be safely or lawfully set aside. All are free agents, and ought therefore to be treated as moral beings, and in no case as animal machines alone. All are immortal, to whom the present life is probation for another life soon to follow; and are to be treated with honour, as called, by Divine invitation, to partake in the Christian hope of that better life to come.

The same principles apply to the still higher duty of those who aspire to frame national constitutions, or to determine the fundamental laws and guiding maxims on which the whole course and tone of national legislation is to depend. When great moral truths are forgotten, and politics degenerate into a mere strife of party, or a series of spasmodic efforts to combine discordant elements into momentary union, and thus to secure at any price the majorities on which power and influence depend, political life must suffer a rapid and sure decline. But I forbear to enter on a subject so wide, which suggests to thoughtful observers grave and anxious reflections on the future of our own country. One thing is sure, as the pillars of heaven and earth. It is not the abundance of private wealth, nor the largeness of the national revenue, which can secure the lasting happiness and welfare of an empire like our own. Our national safety depends on deeper causes, and the true talisman of our lasting prosperity and greatness is of a higher kind. It consists in the supremacy of justice, mercy, and faith in the public counsels of the land; and in the mingled and harmonious culture of the great laws of freedom, order, and brotherhood, sustained and quickened by faith in men's immortality, among the dwellers in countless homes of domestic purity, peace, and love.

LECTURE IX.

MORALS AND REVEALED RELIGION.

THE relation between Moral Philosophy and the Christian Faith, a subject highly important at all times, has acquired new interest in the present day. It has a most intimate connection with nearly all those great philosophical and religious questions, which, in an age of peculiar intellectual activity, occupy the thoughts, and divide the judgment of educated men. Very opposite views have been zealously advocated in former days, and are still maintained. There are Christians of earnest faith, who conceive that all study of Ethics, as a purely human science, whether based on moral intuitions, or on calculations of far-seeing prudence, is set aside and rendered superfluous by the fact of a Divine revelation. All moral speculations, depending on experience and natural reason alone, seem to them nothing better than feeble tapers lighted at noonday.

The exact converse of this view is held by disciples of the idealist school of sceptical philosophy. The clearness and certainty of Ethics, as a science of pure reason, seems to them so complete and self-sufficing as to dispense with all need for supernatural revelation. A religious creed, designed for a moral purpose, and still dependent inseparably on historical records, and needing external confirmation from alleged miracles and prophecies, must therefore, on their principles, be only a useless incumbrance of an ethical system otherwise complete, by which its scientific character is disguised and obscured. It must be little better, in their eyes, than an improbable fiction, an unmeaning superfluity.

Such a view of Moral Science is the basis of Tindal's once famous work, *Christianity as old as Creation*. It has been unfolded, in his own peculiar style, by Immanuel Kant in his essay on *Religion within the bound of Pure Reason*. It has been more lately expounded, in a popular style, with zeal and eloquence, by Theodore Parker, in his *Discourses on Absolute Religion*. It is warmly espoused in a recent work on *Intuitive Morals*, and has plainly found, at the present hour, a wide acceptance among persons of general culture and intelligence, who retain a vague and loose profession of Christian faith. They look upon Christianity as simply the republication of a code of moral duty which had been taught mankind by the light of reason alone. To this extent they will freely allow that it may have, in a practical sense, some real worth and value. But as soon as it attempts to travel beyond this limit, and to set up any claim of a higher kind, to proclaim mysterious doctrines, or remove the veil from the unseen world, they deny its authority altogether. It then lies open, in their opinion, to the serious charge that it encumbers a pure and lofty science with Jewish superstitions and prejudices, histories of doubtful and uncertain truth, a multitude of worthless ceremonies, and dogmas morally useless and hopelessly obscure.

This view of Tindal and other kindred writers was prominently recalled to notice a few years ago, in the sixth of the *Essays and Reviews*, which discussed at some length "*The Tendencies of Religious Thought in England, 1688—1750.*" A real effort to throw light on the difficulties of the subject would have been a most valuable gift to thoughtful and perplexed inquirers into the nature and claims of Christianity. But of such an effort there is no shadow of a trace. The aim seems rather to be, in a series of hasty criticisms and rapid historical sketches, to render the mists which easily gather around a difficult subject still more dense and impenetrable, and the darkness deeper than before. The only conclusion drawn, when a dozen theories have been discredited and set aside, is that no Christian in our days, it is highly probable, has any clear perception on what ground his faith ought really to rest. The historical kaleidoscope is lightly shaken, and a dozen diverging opinions are made to pass before the eye in swift succession, but the youthful reader receives not the slightest help to guide or assist his own judgment. Such a purely literary and curious treatment of the gravest questions of moral duty and religious faith is one of the most worthless and mischievous of occupations. Moral indecision and religious uncertainty are weeds which grow rapidly enough of their own accord, and have no need of any artificial culture, by such negative and purposeless criticisms, to secure an ample and luxurious development in superficial minds. My own aim will be the exact reverse, to unfold on this subject some clear and definite principles, which may serve as waymarks to earnest minds, impatient of mere literary trifling in grave moral questions, who long to feel beneath them a firm foundation of solid truth, on which their faith and conscience, in harmony with enlightened reason, may safely rest.

There are three mistaken views of the relation between Moral Science and Divine Revelation, which must first of all be set aside, before any approach can be made to a just solution of the great problem we are now to consider.

The first of these is the opinion, not unusual with a large number of devout Christians, that Revelation excludes Ethics as a Science, and renders all reasoning on questions of morals needless and presumptuous. The saying of Omar, in the doubtful tradition of Abulpharagius, has found its exact counterpart in some modern reasonings of Christian men. "If these writings of the Greeks agree with the book of God, they are useless, and need not be preserved. If they disagree, they are pernicious, and ought to be destroyed." But such a view, though its advocates may intend, in advocating it, to do honour to the Bible, is one which the Scriptures themselves set aside and condemn. They give no warrant whatever for that blind, superstitious faith, which, in order to establish its own supremacy, and reign without a rival, begins by stifling the voice of natural conscience, and seeks to condemn it to utter silence.

The very first page of Scripture contains a clear proclamation of the moral gifts and faculties of mankind. For man, we are there told, was created at first "in the image of God." Whatever else the words imply, they evidently include the affirmation of those moral capacities, and of that spiritual nature, by which man resembles his Divine Creator, the Holy One who is a Spirit, and is thus widely parted from the beasts of the field. The same truth reappears in another form in the severe expostulation by the prophet: "Woe to them that call evil good, and good evil; that put darkness for light, and light for darkness; that put bitter for sweet, and sweet for bitter." And still later, in the New Testament, our Lord himself, the supreme authority with every Christian,

makes an appeal to his adversaries on this very ground—
"Yea, and why even of your own selves judge ye not that
which is right?" The presence of a moral faculty in man,
which can judge, and ought to judge, even with respect to
the character of a heaven-sent messenger, and the moral
features of a message claiming to be Divine, is thus affirmed
from the opening to the close of the Christian revelation.

Again, the Scriptures teach that there is a standard
of right and wrong, not wholly inaccessible to mankind,
and morally binding on them, even apart from messages
supernaturally revealed. The Gentiles, as St Paul plainly
teaches, were "a law unto themselves," although they had
no law, in the sense of a Divine code, like that of Moses,
externally revealed. And, a thousand years before, the
same truth had been proclaimed by the Psalmist in his
simple and sublime description of the testimony of the
heavens,—" Their line is gone out into all the world, and
their words unto the ends of the earth." The doctrine
which denies the moral nature of man is a vain and
fruitless attempt to silence these heavenly voices. Their
calm and solemn accents never cease, though they may
fall too often on dull and heedless ears.

The Scriptures teach us also, with equal clearness, that
morals are a progressive science. For although, as in
physical science, the truths themselves are permanent, and
do not change, their range is vast and boundless, and there
is growth in the knowledge and apprehension of them by
individual minds. All the terms which we use habitually,
to describe the most real knowledge of sensible objects,
are transferred everywhere in the word of God to this
higher field. It speaks with high praise of those who
make constant advance in a pursuit so high and excellent,
and thus "by reason of use have their senses exercised

to discern good and evil." And a similar analogy, from outward light, and the faculty of vision, is used in the Sermon on the Mount to describe the effect of moral singleness of aim in giving clearness to the understanding, and firmness and consistency to the moral judgment. "If thine eye be single, thy whole body shall be full of light." The communication of fresh gifts and higher privileges is described, by the same Divine authority, as mysteriously linked with the right use and improvement of those already received. "To him that hath shall be given, and he shall have more abundance; but from him that hath not shall be taken away even that he hath."

The principle is also clearly recognized, that moral truths, apprehended by a moral faculty, are one main and essential part of the evidence of a Divine revelation. By this means alone can its reception be fully distinguished from mere credulity and blind superstition. There is an abundant appeal, it is true, to evidence of a lower and more sensible kind. But even here the presence of a moral element is implied. The miracles of Christ were themselves works of mercy, and parables of Divine grace; and the prophecies, to which appeal is made, are described with emphasis as the words of holy men, who spake under the impulse of the Holy Spirit of God. But in other cases this moral element in the testimony stands alone, and appears in fuller relief. "Which of you convinceth me of sin? and if I say the truth, why do ye not believe me?" And the Apostle, treading in the steps of his Divine Master, describes the main object of his own preaching in those impressive words,—"By manifestation of the truth commending ourselves to every man's conscience in the sight of God."

Christian Faith, then, by the testimony of the Scriptures themselves, does not exclude all moral science, or supersede natural conscience, but recognizes its reality, and the duties which flow from it, in the plainest terms. But an opposite error, still more dangerous, must equally be cleared away, before any just view of the whole subject can be attained. For Moral Science, whether under the earlier title of a "Christianity as old as creation," or the modern name of "Absolute Religion," has been held to exclude all need for supernatural revelation, such as Christianity plainly claims to be, and to make it superfluous and unmeaning. By the help of this maxim men have been invited to rise "from the transient shows of time to the permanent substance of religion, from a worship of creeds and empty belief to a worship in Spirit and in Life;" and the claim is made, by this process, to have "removed the rubbish of human inventions from the fair temple of Divine truth."

But the error of this opinion, that the perfection of Moral Science, or of the absolute religion of reason, makes any further and positive revelation superfluous, is shown by a single thoughtful glance at the actual state of mankind. Nowhere in heathen or even in Christian lands are there any signs of such abundance of moral light, of so high and advanced a stage of moral attainment, as plainly to show that any gift of supernatural revelation would be a needless superfluity. The only reasonable objection to the claims of Christianity must be of an opposite kind. It may be urged, with some plausibility, that the moral state of Christendom is so little raised above the level of heathen countries, as to discredit the notion that the gospel is a Divine message, and to countenance the view that it is only one transient and imperfect phase of a purely human development. But clearly the natural conclusion from such a premise can never be that all revelation is superfluous. It must rather be that the light of Christianity itself is too limited, partial, and obscure; and that, even after eighteen centuries during which its influence and teaching have widely prevailed, some fuller and clearer message from heaven is still to be earnestly desired, in order to scatter the still remaining darkness. Now this is the exact reverse of the conclusion which Tindal and his successors have drawn, from the alleged clearness and certainty of ethical truth, to set aside the claim of the gospel to be a Divine message.

This dream of sceptical philosophy is disproved, not only by a direct appeal to the lessons of a long and painful experience, but by a just perception of the higher necessity and importance of moral truth, when compared with the discoveries of physical science. An increase of man's knowledge of the outward laws of nature is desirable, both as a healthy exercise of human intelligence, and for the practical results, promotive of social ease and comfort, to which it leads. Still the knowledge which men require, to provide for the animal necessities of life, is of an elementary and familiar kind, and all beyond seems almost to belong to the class of social luxuries. But uprightness, honesty, and truth, social kindness and good will, the restraint of selfish and angry passions, the culture, if not of high and heroic excellence, at least of temperance, prudence, and sobriety, and the domestic virtues, are almost essential to the peace and welfare of society, besides their connection, which reason shows to be probable, and Christianity proclaims on Divine authority, with the nobler hopes and wider prospects of life beyond the grave. It may be wise and kind in the great Author of our being, when man has once attained such knowledge of nature as enables him to provide for the wants of his animal life, to lay the world before him for ceaseless research, and leave him to discover its laws and harmonies by his own unaided efforts. But if selfishness and passion have clouded the eye of the soul, and make it hard for him to discern higher truths, far more vital to his present and future welfare, the same wisdom must surely point to an opposite course. We may well infer from the Divine goodness, that help for the discovery of moral and spiritual truth will be given to those who sojourn in a land of twilight and shadow, so to assist their feeble efforts to escape from the darkness, and to guide their feet into the way of peace.

This presumption, drawn from the actual wants of mankind, and the urgent need for a practical attainment of moral insight, is confirmed when we reflect on the deep and mysterious nature of these truths, which it is so needful for them to know and understand. Man is higher and nobler than lifeless matter. Human conscience, will, and intelligence, are harder to explore than the facts of physiology, and the laws and instincts of mere animal life. Man, even as he now is, amidst all his follies and vices, is a more difficult study than the depths of ocean or the stars of the sky. But Man, as he ought to be, the knowledge of his high capacities of moral growth and attainment, his duties and dangers, the means and helps of his progress, the noble and perfect standard he should strive to attain, is a subject harder and nobler still. Now this is the great subject of Moral Science. The simple nature of its first elements cannot remove the vast complexity, the unsearchable mystery, which must belong to its full development. It rises from earth, and climbs towards the highest heaven. It passes beyond the bounds of space and time, and reaches into eternity. It strives to reach the mountain-tops of creation, and there still to gaze upward on the unsearchable mysteries of eternal and uncreated love. Some thoughtful minds, when they look on the past and present state of the world, may have doubts and misgivings, that haunt them like dark spectres, whether any direct light from heaven has ever dawned on mankind, or can be reasonably hoped for in time to come. But surely to reject the idea of supernatural revelation on the ground of the clearness of moral and spiritual truth, and the perfect ease of its attainment, so as to render all such help superfluous, is the worst madness of human pride, a climax of unreasoning folly. The grandeur, difficulty, and loftiness of moral truth, no less than the greatness of the moral wants of mankind, form a strong and almost irresistible presumption, that those who are walking in mist and darkness may reasonably hope for some help to clearer vision from a God of infinite wisdom and perfect love.

The same view is confirmed by all the analogies which human parents and teachers supply in every field of human thought. Men are never left to raise themselves out of ignorance by their own unaided efforts. Life begins with a course of education, in which the child passively drinks in the lessons which are set before it by a knowledge higher than its own, and profits by all the stores of truth accumulated by former generations. But moral truths, which involve the duty and the destiny of man, the ends for which the present life is given, and the prospects of another life after death, are subjects in which the best and wisest of mankind, in proportion to their wisdom and loftiness of thought, feel most keenly their own ignorance and weakness. Their highest attainment is the spirit of a little child, which feels how ignorant it is, and longs to be taught and guided by a wisdom far beyond its own. Such an instinct, in thoughtful minds, who feel how steep is the ascent of the mountain-side, and still look upward, is the pledge of its own fulfilment. All human teachers, and the instructions given to children by their own parents, are multiplied proofs, to reflecting men, that our race have not been abandoned, like orphans, to their own self-taught progress in moral wisdom and insight by their Father in heaven.

These reasons, to disprove the strange and wild paradox that revelation is rendered superfluous by the clearness and self-evidence of Moral Science, are further confirmed by the great fact that Christianity, on the very face of it, is a remedial system. It is no message to creatures in a state of moral health, who need no physician. It implies and assumes, on every page, the existence of a dangerous moral disease. It is feet to the lame, and recovery of sight to the blind. It starts from the assumption that men's faculties of moral discernment, though not destroyed, are grievously impaired, and that restoration to health is needful, as well as the provision of external light, before healthy vision can ensue. This view, so plainly taught in the divine message, is confirmed on the largest scale by the past history of the world. The claim, then, that men can dispense with all supernatural revelation, and may safely rely upon their own unaided reason in the whole field of moral inquiry, is like a suggestion that the inmates of a blind asylum may easily build an observatory, provide it with transit instruments and telescopes, with spectroscopes and heliometers, and discover all the magnificent truths of modern astronomical science, without the help of any teacher, oculist, or physician, by their own spontaneous efforts alone.

A third view of the connection between Moral Philosophy and Christian Faith is that which enforces a policy of mutual isolation. Though more usual in its application to Physics, it is often extended to this higher field of thought. Each subject, it holds, must be studied separately, and treated in entire independence of the other. We are to have "no hankering after seeming reconciliations." The geologist, "if loyal to science, will marshal his facts, as if there were no book of Genesis," and the interpreter of Scripture "will explain its meaning, as if there were no science of geology." Attempts to unfold the harmony of revelation and science, even by divines of eminence, have been called "restless, feverish, and sad," and condemned severely, not simply for errors and faults in the execution, but as if the effort, in its own nature, were a criminal and mischievous folly, instead of being an intellectual necessity with every thoughtful and reverent mind.

Such an entire separation, however, in the case of morals still more than of physics, is unnatural and impossible. It could be attained only at the price of two fearful and immense evils, that religion should become immoral, and moral philosophy irreligious and profane. The remark has sometimes been made, that the different source of the two subjects is enough to warrant their entire separation. Human philosophy or science, it is said, depends on experience and induction, but religious faith on Divine authority alone. They ought therefore to be kept entirely apart, and neither of them to be suffered to interfere in the least with the free and full development of the other.

Such expedients, however, to avert all risk of conflict between revealed religion and human science and philosophy, must be wholly unavailing. However good the intention of their advocates, they fight against the fixed laws of human thought, and must utterly fail. We cannot believe contradictions, whatever the sources from which they are derived, or however diverse the evidence or authority on which each rival doctrine is supposed to rest. One of three results must certainly follow. We must renounce the alleged authority of the Divine message, or deny the certainty of the supposed science, or set aside the alleged contradiction, and maintain, on reasonable grounds, that it is illusive and untrue. But we cannot seat our theology in one lobe of the brain, and a philosophy which flatly opposes its teaching in the other. When seeming contrasts arise, we may defer the attempt to reconcile them, believing the materials for a full solution to be insufficient, and may be content to wait for a day of fuller light. But the unavoidable result must be either dimness and uncertainty in our faith in the Scriptures, or some large reserve of doubt and incredulity in accepting conclusions propounded to us in the name of ethical philosophy or natural science.

The whole drift of thought, in modern times, is opposed to the view, in politics or in science, which would find safety, whether for creeds or for countries, in a kind of Japanese isolation. Every where, through all the various fields of human thought, the essential unity of all truth is becoming more and more apparent. The principle is too deeply true to be overlooked and contradicted without loss and danger, whether by Christian believers or by men of science. The latter are bound to receive the evidence of well-attested facts, even if drawn from those fountains of sacred truth which they are sometimes tempted to overlook or despise. The former are bound to serve God with a reasonable faith, and never, in reading those Scriptures which they believe firmly to be the word of God, to neglect the light which must be thrown on their real meaning from all successful and honest study of the works of nature, and from the predicted growth of natural science in these latter days.

Moral Science then, if it be truly a science, must be in full harmony with all genuine revelation. A Divine message, on the other hand, if genuine, must be in harmony with the lessons of true morality. And such is really the relation in which they stand to each other. They are not rivals, but friends. Christianity reinforces lower and plainer truths of morality, and unveils the higher, which are less attainable, and hardly accessible to men's unaided reason. Ethical Science recognizes the moral excellence and purity of the Christian revelation. Whenever their harmony seems obscured, it either serves directly to remove false constructions of a message, so plainly worthy of a Divine origin, in all its main features, or else it teaches the wisdom and duty of distrusting our first impressions, when they would impute, to some particular doctrine or narrative, inwrought into a code so pure and lofty, moral inconsistency and contradiction. The objections in most cases, as in the noted example of Abraham's sacrifice, are due only to the rashness and hastiness of superficial thought, which a juster and deeper view of the genuine conclusions and teaching of Moral Science would be sufficient to expose and condemn, and to reveal the perfect harmony of the sacred narrative with the noblest and most comprehensive maxims of true morality.

These false theories once removed, the way is open for a brief survey of the true connection between genuine Ethics and Christian Theology. Christianity confirms and enforces the main principles of Moral Science. That Science, in its turn, is a needful handmaid, which ministers in various ways to a right reception of Christian faith. Christianity promotes ethical culture, and supplies the largest helps for progress towards the true standard of moral excellence; while Moral Philosophy, in its systematic study, contributes, in many different ways, to the growth and full perfection of Christian Theology.

Christianity, first of all, confirms and enforces the great principles of Ethical Science, because it affirms so clearly the moral element in the constitution of human nature. The first utterance of Scripture, when man is brought on the theatre of sacred history, is to proclaim that he was created in "the image of God." The words are full of the deepest significance. They express the wide interval, which some anatomists strive to bridge over by fixing their eyes on bodily resemblances alone, between all the lower animals, and a being endowed with reason and free will, capable, as a spirit, of communion with the Father of spirits, and of knowing, loving, and serving the Supreme Creator. This keynote, struck so early, resounds through every later page of the Bible. It appears once more in the New Testament, where the main object of the Gospel is said to be the restoration of the lost or defaced image of God to the souls of men. It is this full recognition of man's moral nature, this tone of lofty thought, in which he is viewed as a creature standing apart and alone, made in God's image, lord of this lower world, but subject to a Divine law, and placed under the moral government of the righteous and holy Governor of the universe, which makes the Bible, even where its special doctrines, and the force and drift of its separate messages, are hardly received, or scarcely known, so mighty an instrument, in all ages, for the moral awakening and elevation of mankind. In the light of true morality, its exclusion, in a Christian land, from the outline and course of early education because of the strifes and divisions of Christians themselves, would be at once a folly and a crime. No sadder exhibition could be made of the power of human perverseness to dishonour and cast aside the most precious gifts of Divine love.

Again, the Scriptures reinforce the fundamental truths of Ethical Science, when they insist so strongly, in every part, on the essential contrast between moral good and moral evil. Human probation, in all its forms and varieties, is made to depend on this great and fundamental truth. The commands of the law, the warning voices of the prophets, appeal to it continually. Man is gifted with the knowledge of "good and evil." He is responsible to God for this knowledge, and is bound to use it in choosing good and refusing evil. Life and death are set before him, and he is commanded to choose life, that he and his children may live. A solemn woe is denounced on those, more guilty than the robbers of some earthly inheritance, who seek to remove the moral landmarks, which are the common inheritance of all mankind. A severe invective is pronounced against those who have no knowledge to do good, but are wise to do evil. Our Lord himself appeals to conscience as the eye of the soul, and proclaims the result of its blindness through passion and vice to be the darkening of the whole intellectual being. St Paul teaches that the light of nature, shining in the outward works of God, leaves men without excuse, when they yield themselves to the seductions of sensual vice, and forget or deny the Creator, who has stamped the signs of His own wisdom and greatness on all the works of His hands. Each inspired Apostle, in turn, makes his appeal to this moral faculty in the mind of man. "To him that knoweth to do good, and doeth it not, to him it is sin." St Peter exhorts the first disciples to maintain "a good conscience, that those may be ashamed who falsely accuse their upright conversation." St John lays down the high standard at which the Christian ought to aim : " If our heart condemn us not, then have we confidence before God." And St Paul sums up the whole outline of moral duty in that comprehensive appeal: "Whatsoever things are true, whatsoever things are honest, whatsoever things are just, whatsoever things are pure, whatsoever things are lovely, whatsoever things are of good report; if there be any virtue, and if there be any praise, think on these things."

But further, the Bible confirms and establishes the foundations of Moral Science, because it teaches plainly that the standard of right and wrong is no arbitrary thing, which comes into being through positive appointment, or some changeable act of the Divine will, but a law which is true before its own revelation, and on this account has been revealed. It does not fear to refer the acts of the Almighty to this standard, and thus teaches that Eternal Goodness is a deeper and holier attribute than even Almighty Power or Omniscient Wisdom. Its voices by the earliest patriarchs are these: "Shall not the Judge of all the earth do right?" "Shall he that hateth right govern, and wilt thou condemn the Most Just?" It sets before us the Supreme Creator making this appeal to His own creatures: " Are not my ways equal? are not your ways unequal?" It crowns the messages of the law with that most glorious of all moral aphorisms with regard to the stainless perfection of the Divine goodness: " He is the Rock, His work is perfect, all His ways are judgment; a God of truth and without iniquity, just and right is He." And in its later messages it sets before us, as the highest object of Christian desire, to make trial, in our own conduct and course of life, what is "the good, and perfect, and acceptable will of God."

Another way, in which the Scriptures confirm the reality, and enforce the authority of Moral Science, is by laying it down for a fixed law of Christian duty to make progress continually in spiritual discernment, and in the apprehension of spiritual truth. No countenance is there given to the crude fancy that Ethics, viewed on their practical side, are a barren and unprogressive subject of thought. We rise into a purer atmosphere, where the mephitic choke-damp of such depressing theories finds no place, but is swept away by the fresh wind of some holier influence, and the human conscience breathes freely once more. The voices it utters are full of reality and of hope. They encourage, while they stimulate, the willing learner to climb higher and higher up the hill-side of truth. "To him that hath shall be given, and he shall have more abundance." "The path of the just is as the shining light, which shineth more and more unto the perfect day." The soul, we are taught, has senses resembling those of the body, only more wonderful and mysterious. And these, like their outward counterparts, grow keener and finer in their perception, when rightly employed, so that Christians, "by reason of use", may attain a fuller and clearer knowledge, both of good and evil. Such a growth of knowledge, in the case of unreal shadows, is plainly impossible. The reality of Moral Science is plainly implied, when such various degrees of moral discernment lie open to the attainment of the sincere and humble Christian.

But the New Testament supplies a still deeper confirmation of the same truth by the description which it gives of the main drift and aim of its own message. What is the great end, to which miracles and prophecies, the ordinances of the Christian church, and all the secondary features of the revelation contained in the Scriptures, are subservient? What is the standard of hope and desire, which its disciples are taught ever to keep in view? It is not blind submission to the decrees of almighty power, or the dull, unreasoning apathy of the fatalist, but the full, intelligent harmony of the created will with the felt and known perfection of the Divine goodness. Its highest message is not couched in the sentence that "God is Power," but that "God is Light" and that "God is Love." It teaches us to see a moral goodness, perfect and without stain, seated on the throne of the universe. Towards this high and perfect standard it represents it as the privilege and duty of the Christian continually to aspire. And thus it leads us to rise above the domain of natural sequence and mere physical law, in which the ebb and flow of material changes is subject to rules of quantity and direction alone; till we reach a higher region, where the intelligent will, gazing on a perfect standard of Divine wisdom and love, approaches nearer and nearer to the glorious object on which its gaze is fixed unceasingly, shines more and more with some real though faint reflection of its Divine beauty and glory, and sees light in the light of heaven. The character of the Christian revelation is wholly misconceived, it is true, when it is taken merely for the republication of a code of morals, in which no remedies are provided for the moral disease of mankind, and no powerful helps are given, and no mighty attractions of Divine grace are revealed, to replace the barren acknowledgment of a law perpetually transgressed by the peace of forgiveness, and conscious progress towards the recovery of the lost image of God. But still it is certain that the aim and purpose of the gospel is the moral elevation and spiritual recovery of those to whom its message is given. And no surer pledge could be given to us of the unspeakable dignity and importance of Moral Truth, when contrasted with mere physical progress, than a just view of that wonderful economy of Divine grace which the gospel reveals. Its final aim, as clearly announced by prophets and apostles, is to plant the laws of duty once more in the hearts of men, so as to provide an effectual remedy for the moral disease which has spread so wide, and lasted so long; and in due season to establish a kingdom of purity and love, of righteousness, peace, and joy, which may form an image and reflection on earth of the moral beauty and perfection of a higher and brighter world.

LECTURE X.

MORALS AND REVEALED RELIGION.

THE reality and excellence of Moral Truth, viewed as a science and object of human thought and study, is confirmed, and not contradicted or impaired, by Christianity. But the relation of the two subjects is mutual, as well as intimate, and Morality has an important office to fulfil in connection with the right reception of any message which claims to be Divine.

How may we discern between a true and a pretended revelation? The credentials of such a message must answer to the perfections of its Author. They are Divine power, far-seeing wisdom, and supreme goodness. All these attributes, though perhaps in various measure, may be expected to leave their impress on every oral utterance or sacred writing, which really comes from God. The moral evidence cannot usually stand alone. In that case the reception of a heavenly gift would be made to depend on the perfect and healthy working of a faculty, grievously imperfect and diseased, which creates, by its disease, the need for this Divine remedy. But though moral evidence alone is not enough, because of the state of those to whom the message is given, it is one main element in the credentials of any true revelation. Its presence is essential, in order that the acceptance of the message, as really divine, may be an act of genuine faith, and not of blind credulity. The unholy and impure must be quite unable to pass a true moral judgment, in detail, on every part of a record which claims to be Divine. Still, if it be otherwise well attested, they may be perfectly able to judge whether its general tone is worthy of a message from the God of heaven; or whether, from its low tone, and confused moral teaching, it is wholly unworthy to be received in the high character of a genuine revelation of His will.

Ethical Science, again, prepares the way for Christian Faith, because it acts as a powerful detective to disclose the moral wants of mankind. All spurious moral teaching lowers and debases the standard of right and wrong, so as to quiet the uneasy conscience, and bring down the rule of duty to the level of the actual state and conduct of men. But true morality teaches a widely different lesson. It refuses to lower its standard, in order to relieve its subjects from the painful consciousness of wrong-doing. It brings out into clear relief the wide and deep contrast between the laws of right, and the actual state and history of the world. And thus it reveals on the side of science, and seals by the voice of natural conscience, that lesson of the wide and universal spread of moral evil among men, which forms so main a feature of Scriptural Theology. It re-echoes, in tones of power, those heavenly voices, which announce to mankind the presence of sore disease, and their need of some effectual and Divine remedy.

Again, Moral Science, when pure and genuine, includes and recognizes those deep instincts of creature dependence and conscious weakness, which, in a race like our own, render a Divine revelation the natural object of hope and earnest desire. The full soul loathes the honeycomb.

It is possible for those who have been reared amidst the light of Christian teaching, and who are still averse to the high and pure morality of the gospel, to turn away with a cold indifference from whatever appeals to them in the name of God, and claims to be Divine. The state of mind which proclaims all knowledge of the Creator impossible to be attained, and then proceeds, without one pang of regret or sigh of bitter lamentation, to frame its own theories of a universe consigned to eternal and hopeless darkness, is one which finds unhappy illustrations in our own day. But the moral instinct, when really awake and in vigorous exercise, and moral science, when its voice is truly heard, teaches a widely different lesson. Its voice is that of the ancient patriarch in his hour of sorrow, "O that I knew where I might find Him! that I might come even to His seat." Even in heathen lands it will prompt a secret yearning of the spirits of men, to "feel after the Lord, if haply they may find him." It imprints on every thoughtful mind a deep conviction that contentment with entire religious ignorance and uncertainty, far from being the sign and test of advancing science, is the sure mark of a deep and ineffable moral degradation. There are only two states of mind, which, in the light of sound ethical teaching, can be safe or right in a creature like man, gifted with such high powers, and still conscious of their past abuse—to know and love God who has been actually revealed, or to seek with earnest longing of desire after the God who is still unknown.

Right and sound moral teaching, when a divine message has been outwardly received, will aid in the work of setting aside false interpretations, to which it must always be exposed, and which tend to obscure or pervert its meaning, and thus to defeat the object for which it was given. The best and noblest gifts of God are liable to serious abuse, when entrusted to the keeping of sinful men. The Scriptures may be wrested, as an Apostle has warned us, by men of perverse mind to their own destruction. And even where the abuse is not so extreme, genuine faith and piety may be mingled with no small amount of superstitious ignorance or blind credulity. Under the plea of reverence for the word of God, the faults of good men may be mistaken for virtues, holy words may be wrested from their context, and violently forced into the mould of imperfect human systems, or supplemented with human guesses and fancies, till their true meaning is obscured or wholly forgotten. Here a genuine knowledge of ethical truth may prove an important safeguard against serious dangers. It will teach men that records of human error and sin, even when they are found in a Divine message, do not suffer a moral transformation, and become examples of moral excellence ; and that special duties, depending wholly on some special command of God, are no laws for universal guidance. Actions, which would be highly criminal, when their motive is the mere caprice of self-will, or the excitement of passion or revenge, may sometimes be proofs of heroic faith or earnest love, when done in obedience to the express command of God, to fulfil some great end of His moral government. And outward and ceremonial precepts, in which the principles of moral duty are mingled with details, and applied to variable and transient circumstances, may be set aside by the Lawgiver who appointed them, and still His wisdom may be perfect alike in their earlier institution and in their later abrogation.

Again, Moral Science, rightly understood, is the spouse and ally of Christian Faith, because it reveals and enforces the supreme importance of that knowledge of Divine truth, and of the character and perfections of the Creator, which it is one main purpose of the gospel to impart to mankind.

The doctrine, lately taught by some eminent writers, that nothing can be known of God and His moral nature, because He is an infinite Being, is directly opposed to the whole scope and aim of the Christian revelation. Its effect, whenever consistently held, must be to destroy all Theology and all Ethical Science at one common blow. In the hasty recoil from speculative theories of religion, the rivals and substitutes of Christian faith, it would plunge the world and the whole church into a gulf of hopeless darkness. But the view is not more opposed to the teaching of Scripture than to the voice of conscience and sound reason. All truth is so closely linked together, that a fatal necessity of entire ignorance in any one field of thought must extend its influence, like a mist of gloom and obscurity, to all the rest. If we know nothing at all on any subject of which our knowledge is not exhaustive or complete, no person or thing in the wide universe can ever be really known.

Moral Science, when its truths are clearly and firmly held, teaches plainly the opposite lesson. It thus becomes the natural ally and handmaid of Christian faith. We do not know ourselves perfectly, either in body or soul; and in this life, perhaps even in the next, we can never fathom all the mysteries of our own being. Yet we know that we have an inward power of choice, and are bound to use it aright. A light shines down upon us, which bids us seek continually a perfection not yet attained. We do not know perfectly our nearest relative or dearest friend. The social ideas, on which all human fellowship is based, are mysterious and obscure. Yet we know and are sure that kindness and love to all, but especially to neighbours and friends, are our bounden duty and high privilege. Our ignorance of the secret heart of others, and of many hidden laws on which their health and comfort depend, cannot set aside the great maxim of the Apostle, "To him that knoweth to do good, and doeth it not, to him it is sin."

The same principle must apply when we travel upward in thought to a higher sphere. The greatness of the Creator is unsearchable. The Infinite Being transcends, and must transcend, the full comprehension of every finite understanding. The voice of the ancient patriarch is no recent discovery of an improved philosophy in modern times : " Canst thou by searching find out God ? canst thou find out the Almighty to perfection?" But when a fatal necessity of entire and absolute nescience is inferred from the mystery, which hangs around the horizon in every field of human thought, and is then applied to religious truth alone, the reason, conscience, and heart of man protest against a conclusion no less illogical and baseless than debasing and repulsive in its moral aspect.

There are many dark lines in the solar spectrum. It shares this feature, though in more abundant measure from its superior brightness, with flames of earthly origin and inferior brilliance. Yet the spectrum is light, and not darkness. These lines, though dark themselves, could never have been discovered by those who were wholly blind. And the same principle clearly applies to a higher subject. The very power to perceive and know the mysteriousness of the Divine nature is a proof that some partial knowledge is attainable. Minds which can speculate on the persistency of force in all nature amid cycles of perpetual change, on the secret undulations of light, and the mass and velocity of stars almost infinitely remote, have certainly powers by which they can know, and ought to know, their Almighty Creator, and thus to echo the voice of the Psalmist : " Be sure that the Lord, He is God ; it is He that hath made us, and not we ourselves." They may feel, and ought to feel, the certain truth of that moral axiom of the Apostle : " The invisible things of Him, from the creation of the world, are clearly seen, being understood by the things that are made, even His eternal Power and Godhead." And from this great fact, as its fountain, another truth must naturally flow, that contented ignorance of that God, in whose hand our breath is, must be always and in all circumstances a grievous and inexcusable folly. It is a threefold offence, against the true dignity of man as a responsible creature, against the bonds by which all human society is held together, and against the authority, love, and kindness, of our Father in heaven.

Christian Theology, in many different ways, confirms and deepens the claim of moral truth on the minds and consciences of men. It republishes the great facts of man's moral nature, and of his responsibility to his Maker and Preserver, and proclaims them anew with Divine authority. By the voice of that law of holy love which it clearly reveals, it awakens the conscience from its frequent slumber, and brings to light once more the secret characters once engraven there by the finger of God, but since obscured, as in some palimpsest, by worldly habits and selfish passions. It rescues men from that palsy of hopeless despair, which is liable to seize on them, when once aroused from the stupor of moral indifference, and when they see how far above their actual state is the standard they ought to have fulfilled, of which the obligation can never cease, and feel how unable they are to attain it by their own feeble efforts. By new and higher motives, drawn from the revealed facts of the gospel, it calls their moral faculties once more into vigorous exercise, animates them with new grounds of hope, and stirs the heart with a noble ambition to seek for glory, honour, and immortality in the paths of humility, obedience, faith, and Christian love. It provides tests, by which to discover moral progress or decline, and sacred landing-places in the ascent of the soul from lower to higher states in the knowledge and discernment of spiritual truth, and its practical application to the guidance of the life and the culture of the heart. The beatitudes, for instance, are like a ladder of moral ascent and progress, of which the foot rests upon the earth in the confession of human weakness, but the top reaches to heaven, where the soul attains to a full resemblance of the Divine perfection. And besides its other characters, which make it the most powerful instrument of ethical progress, it is one great excellency of the Christian revelation, that it provides, in every part of its inspired and holy messages, such large and ample materials for the illustration and practical enforcement of almost every variety of moral truth.

But the converse is equally true, and seems to claim a fuller development, so as to remove the suspicion which some Christians have entertained, that Moral Philosophy in its systematic study is a dangerous rival and adversary of Christian faith. Such a view, however it may be palliated by the unbelieving tone of some ethical speculations, is highly unreasonable. Its acceptance, by any number of religious men, would be a dangerous stumbling-block in the way of simple minds. For indeed Moral Philosophy, when true and genuine, must be the spouse and ally of Christian faith, and minister in various ways to its progress and full perfection.

First of all, sound ethical teaching confirms and ratifies the grand outlines of revealed morality. It discerns in them great, fundamental, unalterable truths, applicable to mankind in every age, and in all the varying circumstances of human history, and worthy to be enforced anew by direct messages from heaven, supernaturally revealed.

It has been remarked, in a recent defence of utilitarian ethics, that the morality of consequences, to be true and sound, must include not only Epicurean, but many Stoic as well as Christian elements. Such a statement serves only to disguise the real truth. Three main elements really enter into all moral questions, and ethical systems are distinguished and contrasted by the way in which they deal with them. The first is a law or rule of right and wrong, going before; the second, the moral faculty or capacity of moral feeling and judgment, by which that law is apprehended; and the third is the discernment and anticipation of results that will practically follow from morally right or wrong conduct. The transcendental and ideal philosophy of modern times relies on the first alone, and attempts to soar out of the region of common sense and practical expediency into one of absolute, inflexible reason. The old Stoic morality of nature, and its modern counterpart, the ethics of conscience and the moral sense, dwells exclusively on the second element, or the harmony of human action with the voice of personal conscience, as a faculty properly and rightly supreme. The morality of consequences, whether in its older selfish and Epicurean form, or modified by a tacit adoption of the law of universal philanthropy, dwells on results alone. But Christian Morality includes all three elements alike, and assigns to each its due place. It exhibits, first of all, a law of essential right, which it is the duty of man to learn and apprehend for his own guidance. It recognizes, next, a faculty of moral discernment, a conscience, the eye of the soul, which needs to be kept single and pure, and neither warped by pride nor selfishness, that the whole body may be full of light. It dwells further on the truth, that the consequences of actions, soon or late, must answer to their moral character; so that, by an inverse process, their good or evil results will serve in most cases to illustrate and confirm that ethical contrast on which they secretly depend. Thus the genuine teaching of Moral Science, and the voice of Christianity, will be found, the more closely they are examined, completely to agree.

The Bible needs to adopt no foreign element, whether Stoic or Epicurean, to complete its own ethical teaching. It contains within itself whatever is true and sound in either system, as well as in modern transcendentalism, but combines their divergent and imperfect theories in a higher unity of balanced and perfect truth. On the other hand, while it includes the elements necessary to a complete and harmonious system, it does not attempt to exhibit them in a systematic form. And thus the genuine progress of Moral Science must fulfil, towards the ethical precepts of revealed religion, a double office. It will reveal more and more the truth and consistency of those principles which underlie the whole range of Bible morality, in its high appeal to laws which are "settled in heaven," its direct challenge to the human conscience, and its constant reference to practical consequences, in life or death in blessings and curses, that follow on well or illdoing. But it will also develop those principles in other directions, which do not lie within the special aim of the Christian message. For as the Scripture records sunrise and sunset, but does not anticipate the discoveries of Astronomical Science, so its enforcement of great moral truths, and of their sure consequences both in this life and the life to come, leaves still a wide and ample range for the reasonings and conclusions of Ethical Science, when it traces out more fully the first principles themselves, and applies them in fuller detail to all the countless varieties of human character, and the vast and ever multiplying diversities in the complex economy of human life.

Moral Science, when its first lessons are clearly apprehended, performs another service to Christian truth. It establishes, on grounds which are unassailable, the mournful contrast between the true ideal standard of right conduct and the actual state of mankind. And thus it bears witness to that great fact, which is also the fundamental doctrine of the Gospel, that the state of men is one of sin, guilt, and unholiness, of wide departure from the true and inflexible standard of the good and right, and therefore calls for the intervention of some Divine remedy. Every form of spurious morality strives to lower the standard, in order to get rid of this mystery of evil, so solemn and mournful, and to reduce its own requirements nearly to the level of the habitual practice of mankind. The lowest varieties dispense with the law of duty altogether, and make prudential selfishness the only rule of human conduct, subject to no other penalty than the natural results of a defective calculation. In this way, however, not only the harmony of Morals and Christian Faith is at an end, and replaced by direct opposition, but Ethics, as a science, having definite objects and laws of its own, is wholly destroyed. It becomes a mere branch of arithmetic. Its problems retain in this case no moral feature whatever. They merely propose for solution certain equations of a very high order, involving countless coefficients, in which not only the quantity sought, but nearly all the coefficients by which its value is to be determined, are equally unknown.

This harmony, however, between genuine Ethics and the truths of Christianity, is not confined to the mere discovery of widespread moral disease, which proves the desirableness of some supernatural provision for its cure, and opens the way for the disclosure of a remedial system of grace and a Divine Physician. It extends further, to the Incarnation, one great central doctrine of the Christian faith.

There is much in the statement of the Gospel, that "the Word was made flesh," and that the Supreme Creator has been manifested in human flesh, to stagger the faith even of devout men, and to awaken the active opposition and incredulous disbelief of mankind, when we view it on the side of experience or natural probabilities alone. The question of Solomon—"Will God in very deed dwell with men?" presents itself here to the reflecting mind in a startling and almost oppressive form. It can be no matter for surprise that multitudes, not wholly incredulous, shrink from a doctrine which involves so deep a mystery, and feel an utter disproportion between the littleness of man and the place of our planet in the universe, and the alleged union of the Divine nature with our own in one historical Person, who has lived and died, like a man, among his brethren of mankind.

But here the teaching of Moral Science comes to the

aid of Christian Faith, and throws an immense weight into the same scale, when loaded so heavily with doubts and objections on the opposite side. For it teaches clearly two main truths, which go far to confirm the doctrine, so startling at first sight, and still so clearly announced in the Scriptures. First, it assures us of the doctrine, the secret foundation of all genuine Ethics, that the great First Cause, the Creator and Preserver of the universe, is a Being possessed of all moral perfection, a God of spotless holiness and perfect love. And then it assures us, further, that infinite condescension must be one integral and inseparable element in our conception of all-perfect goodness. It is true that the forms of this condescension, and the special features it may assume, or the moral limits that regulate its exercise when it encounters sin and evil in its objects, are too high and hard a problem to be solved beforehand by the guesses of human wisdom alone. Such knowledge is too wonderful for our powers of discovery. But when revealed in another way, the truth may be seen at once to harmonize with the deepest and purest lessons of Ethical Science. For goodness does not and cannot consist in selfish pride, or in haughty isolation. True dignity does not depend on empty state and outward show, but on moral excellence alone. Genuine love delights in stooping to want and sorrow. Supreme perfection can belong only to a Being supremely gracious, and in its modes of disclosing its Divine preeminence may be expected to surpass the dim anticipations of our narrow minds. It is morally worthy of the Most High to "have respect unto the lowly;" and the starry heavens, while they reveal the glory of the Maker, only add new emphasis to the moral beauty of the truth, of which the Incarnation is the highest proof and illustration, that the Lord of all worlds is mindful, with most tender love, of all the creatures of his hand.

The "unknown God" whom the Athenians ignorantly worshipped, was no other than the Father of spirits, who has revealed himself under the gospel in the person of Jesus Christ. But "the unknowable God," which some schools of modern thought offer for our acceptance under that name, is widely different. A Being, of whom nothing whatever can be known, who abides apart, like the gods of Epicurus, in solitary isolation and inaccessible mystery, never to be broken by one gleam of heavenly light, is something worse than a mere negation and intellectual void. If the great mystery of Christian faith overwhelms and dazzles us by the bright vision it presents of an infinite condescension, surpassing our largest efforts of thought to conceive its greatness, this negative theology, which pretends to replace it, is incredible for an opposite reason. It offers to us a moral paradox of the most prodigious kind. A Being of unlimited power, who still is unable or unwilling to have communion with any finite intelligence, who shrouds himself in a mystery never to be broken by one gleam of light, and leaves the world to run in one vast cycle of wearisome and perpetual change, abiding in the selfishness of an eternal isolation, can be no object of honour and esteem, much less of loving awe and holy adoration. It is a spectral vision of darkness, the moral incubus and nightmare of a deserted and desolate universe. Physical Science disclaims the doctrine as an idle, inconceivable figment. But Moral Science goes further, and proclaims this apotheosis of perfect and eternal selfishness to be hideous and impossible.

There is, in reality, no main element of Christianity, which does not find some direct support and confirmation of its truth in the lessons of genuine Ethical Science. For this discovers, in the light of a long and mournful experience, the moral feebleness and impotence of man, the wide extent of corrupt desire and wayward and perverse action. And thus it shows the suitableness to the wants of human nature of that great promise of the Divine Sanctifier, the Lord and Giver of spiritual life, which constitutes one main part of the Christian message. It rescues the doctrine from those intellectual subtleties, by which it has been overlaid and disguised in the many controversies of the church, and presents it, in all its Divine simplicity, as a truth in full and deep harmony with the known wants of the human race, confirmed by the history of the world in every age. It cannot of itself resolve the hard problem, how a Being of perfect goodness and infinite condescension may be expected to deal with a race so morally helpless, and even perverse, as mankind in general are proved, by long experience, to be. But it points our thoughts in that exact direction, where the sunrise dawns upon us in the gospel, revealing the fact of a Divine Comforter and Sanctifier, provided by the goodness of a heavenly Father for the urgent moral wants and necessities of mankind. It confirms, by the voice of natural conscience, gaining fresh light in proportion as it meditates more closely on the laws of duty, the revealed warnings of a judgment to come. It discloses further the complete harmony between the deepest instincts of our moral being, when once thoroughly awakened and aroused, with its dim longing after a moral perfection seen in the far distance and not attained, and the great and glorious hope, revealed in the gospel, and assured to those who receive its message aright, of a perfect and sinless life beyond the grave. As in water face answers to face, so do the natural instincts of the human heart respond to this voice of Divine revelation, and echo not only its warnings of future judgment, but its hope of a higher life and purer moral excellence, to be attained hereafter by all who are true and upright in heart.

Once more, Moral Science vindicates the Bible from a large class of objections, and helps to remove a fertile source of objections to its authority, while it points out, in the fact of a series of progressive revelations, a wise correspondence with the actual wants of the human race.

Wise expediency is not the sole duty of man. It is only one aspect, and neither the highest nor the noblest, of Moral Truth. But it is an aspect inseparable from the rest, and it claims the fullest prominence in the case of legislation, whether human or divine. For the main object of positive laws is to supply the want which arises from the shortsightedness of men in general, and their vice, perverseness, or moral weakness, by the higher wisdom and foresight of the lawgiver, and his freedom from the disturbing influence of mere self-interest and passion. When moral laws have to be applied to the varying circumstances of mankind from age to age, variable elements must enter into their wise application. It is thus a natural and certain corollary from the first principles of moral truth, that a revelation, to be wise, must be moulded, in part, by the actual state of those to whom it is addressed; and that, while the principles on which it is based must be firm and unchangeable as the pillars of heaven, successive modifications in the outward form of the message, far from obscuring and contradicting, only tend to confirm and establish its Divine origin.

There is no reason, then, why firm believers in Christianity should look with a jealous and unfriendly eye on the earnest cultivation of Ethical Science, and still less why the students of that science should turn aside in sceptical doubt, and with secret repugnance, from the messages of Holy Scripture. The distinction between Moral Philosophy and Christian Theology is important, and needs ever to be kept in view. But the elements of union between them are more important and vital than the partial contrasts. In the nineteenth Psalm, long ago, we have a Divine prophecy of their perfect reunion. All the voices of nature, whether of the stars in the firmament, or of natural conscience in the human heart, whenever their sound goes forth to the whole world, tend to resolve themselves into a still nobler utterance, and to lose themselves in a clearer and brighter light, the written word of God. For this proceeds originally from the same source, but, when rightly welcomed, approaches nearer and nearer in its progress to the full and perfect daylight of the better life to come.

LECTURE XI.

ON ETERNAL AND IMMUTABLE MORALITY.

FROM the spiritual geography of Moral Science, or its relation to the various branches of knowledge which compose the great cycle of human thought, I proceed now to consider separately some of those main elements on which its truth and authority depend. It includes three different aspects, the source and ground, when parted from each other, of conflicting ethical systems, and on the union of which, in their due order and proportion, Moral Philosophy must be founded, if these debates are to be reconciled in a deeper harmony. The first and highest is the objective aspect of Morals, which recognizes an eternal law of right and wrong, prior to the perceptions of the individual conscience, and independent of its errors; a lofty and inflexible rule of righteousness, uncreated and everlasting, based on the essential relations of moral agents to each other and to their Creator, an ideal standard of perfect goodness in all its forms, whose seat is the bosom of God, and her voice the harmony of the world. The second is the subjective aspect, which recognizes conscience as an inward faculty of the soul, naturally claiming precedence and supremacy over all the rest. It is this great fact, which makes man "a law unto himself," bears witness to the claims of duty, and awakens inward approval and disapproval on the contemplation of virtuous or vicious actions in ourselves and others. The third aspect of Morals is that which unfolds the connection between right and wrong conduct, good or evil actions, and the wide-spreading consequences that must naturally ensue. It is the great fact which forms the basis of utilitarian theories of Ethical Philosophy. However defective and dangerous, when laborious calculations of pain and pleasure are used to set aside the law of God without, and the direct voice and authority of conscience within, it is an element of moral reasoning which can never be overlooked without serious peril on the other side. A practical code, tested by experience, and suited for the real conduct of life, would then be likely to be displaced by 'some high-sounding and pretentious, but thin, abstract, and barren theory.

Eternal and Immutable Morality is the title of a well-known work of Dr Cudworth, a contemporary of Dr Knightbridge, and one of the chief lights of Cambridge two centuries ago. The objective certainty of moral truth is there maintained with much ability and learning. But the long delay of the publication, which took place forty years after the author's death, when a new school of thought had set in, and still more its scholastic style, have conspired to rob it of its due influence. It needs pains and effort to separate its solid ore of thought from the learned dust and sand in which it is almost buried. Dr Whewell has remarked that "though always mentioned as one of our standard works on Morals, it produces little impression on most of those who view it as an ethical work." One reason he assigns is that its principles are not made to bear manifestly on moral questions, and are too abstract and general in their form.

that the chief cause for the want of any wide influence is the scholastic mould into which many of the thoughts are cast. Such a passage as the following, though a really important and vital truth lies hid beneath its surface, is not likely to secure a wide school of disciples in the present day.

"It is a thing we shall easily demonstrate that moral good and evil, just and unjust, honest and dishonest, if they be not mere words without any significance, cannot possibly be arbitrary things, made by will without nature; because it is universally true that things are what they are, not by will, but by nature. As for example, things are white by whiteness, and black by blackness, triangular by triangularity, and round by rotundity, like by likeness, and equal by equality. Neither can Omnipotence itself, to speak with reverence, by mere will make a thing white or black without whiteness or blackness, or by mere will make a body triangular without having the nature and properties of a triangle in it; nor circular without the nature of a circle, that is, a circumference equidistant everywhere from the centre. The reason is because all these things imply a manifest contradiction, that things should be what they are not."

To place the subject, if possible, in a clearer and more convincing light, it may be well to consider, first, in what sense moral truth must be owned to be contingent, and neither immutable nor eternal; and next, in what sense these high titles of honour may be fitly applied to it, and express a view of its nature, essential to its vital influence and practical power over the human heart.

Now here it is needful, first of all, to distinguish three main divisions of Ethical Science, the morals of desire or affection, of knowledge or intelligence, and of outward action. The first of these is the subject of Pure, the third of Applied Ethics; while the second, which deals with the right conduct of the understanding in all moral subjects, holds a kind of middle place, and serves to form the transition from pure moral truth to its outward applications.

Even Pure Ethics, or the morals of internal affection and desire, involve plainly some contingent elements. They assume the existence of moral agents, with their power of choice, and mental attributes of will and desire. No laws can be in actual force, where there are no persons to whom they apply. The absolute and perfect goodness of God, the secret fountain on which all the streams of moral duty depend, and from which they flow, is so distinct in kind, and so mysterious, as scarcely to be included in the general conception of its nature. For this requires a superior, whether some person, or only a law, to which obedience is due. Now all created being is and must be contingent. Hence Morals, so far as they assume the existence of moral agents, bound by ethical laws, must be contingent also.

Morality assumes the existence, not of the moral agent alone, but also of other beings, towards whom he lies under various obligations. In all social ethics the objects of duty, as well as the agent himself, have only a contingent and not a necessary being. And thus the eternity and immutability of all moral truth of a social kind must be doubly limited, from the mutable, dependent nature of the subjects in whom it inheres, and of the objects to whom it belongs.

In the morality of the intellect other elements of contingency appear. For this class of moral duties depends on the existence of powers and capacities in the human mind, by which the good and desirable may be known, and on the actual limitations in our faculties, which involve the need for slow and gradual progress in the acquisition of truth of every kind. It includes further all the special conditions in the means available for the increase of knowledge, which may vary widely, according to diversities in natural gifts, or in the states of society. All these elements may introduce some corresponding diversity into the actual laws and limits of moral obligation. Truths, fixed in themselves, may thus be elastic and contingent in their application to real life. How much every one is bound to learn and know, even in moral questions, will depend in part on his own natural powers of thought, and in part on the variable means of progress and improvement which he may have enjoyed. The lesson, taught by an Apostle, clearly affirms this great principle of Ethical Science, which links human duty with the opportunities of knowledge previously given: "To him that knoweth to do good, and doeth it not, to him it is sin."

In Applied Ethics the element of contingency enters in still larger measure. Here diversities arise from all the varieties of actual power in every moral agent. The larger his own gifts, the fuller will be his responsibility. All the differences which exist in the capacity of sentient beings for pleasure and pain, all diverse experiences of good and evil in our fellow-creatures, multiply the possible variations in the practical rule of right or wrong action. The great law of love mounts upward, from the lower animals to men, from subjects to rulers, from men to angels, from creatures to God. It includes the whole range of sentient being. But in every stage of ascent, short of the highest, it must include many contingent and visible elements, depending on the actual gifts of our fellow-creatures, and the present constitution of the world in which we live.

No universe, whether natural or moral, can subsist on geometrical or necessary truths alone. Some positive law seems essential to the nature of all created being. Such are plainly the laws of attraction and cohesion in the outward and material world. In the higher field of moral activity, there need also to be positive laws of Divine appointment, which are not incapable, in their own nature, of reversal or change. Actions may be linked with certain results in a manner which all experience confirms, but which we are unable to trace to any abstract or fatal necessity. These laws may be of two kinds. Some of them may refer to the human body, and to the construction of the material world. Others are mental, and refer to the laws and associations of human thought. The murderer, who plunges a dagger into the heart of his victim, commits a hateful crime. But the essence of that crime consists in the malice and hatred, which knows that life is precious, and intends to destroy it. That this outward act should imply deep malice depends on those laws of human life which experience has revealed, and on the murderer's knowledge of the preciousness of life, and of the means by which it can be irreparably destroyed.

Morality, then, cannot be eternal and immutable, so far as it must depend, practically, on a large variety of contingent and variable elements. The positive laws which the Creator has appointed, and which enter so largely into the whole economy of human life, must have a great influence on the application of ethical rules and maxims, however deep and firm their foundations may be.

Still, when every needful concession has been made, there is a sense, deeply important, in which we may claim for moral truth, in the words of Cudworth, that it is eternal and immutable. Even in geometry, the most fixed and certain of human sciences, we must begin by assuming the existence of an outward and spatial world known to us by experience. When this assumption has been made, the relations of geometrical science necessarily follow. They are discerned as necessary, and not contingent, by the general consent of thoughtful and educated minds. Even those philosophical heretics, who strive to resolve them into arbitrary inductions of a partial experience, capable of reversal, will usually be found, in the course of their reasonings, to bear unconscious testimony to the truth they seek to disprove.

The nature of Moral Science is, in this respect, precisely the same. To prove the existence of a moral world, or of beings possessed of will and the power of choice, we must refer back to our own consciousness, and to the experience of others gradually acquired in the course of life. But when this first step has been taken, the power of choice and the sense of moral responsibility are found to be inseparable. In strict agreement with the moral faculties, as soon as their existence is recognized, thoughtful minds are compelled secretly to own that there is some standard of duty or transgression, of right and wrong, binding in its own nature on every moral agent. That standard, embodied in its simplest expression, is threefold, in reference to the threefold faculties of the reasonable mind. It is love in the desires and affections, truth in the understanding, and energy in the will. And the passions or evils which are the contrasts of these, are self-condemned; in the affections, selfishness and malice; in the understanding, ignorance, error, and delusion; and idleness and violence in the field of outward action.

This great truth, however simple in its own nature, is capable of a large and various development. It extends to an immense variety of objects through the whole range of the universe of being. It includes the Creator and all His creatures, angels and unseen spirits, as well as the human race. It comprehends all the various and complicated relations of human life; and even the lower creatures, so far as they are capable of pain and pleasure, cannot be exempted from the circle of its just authority. It is modified by all the immense variety of sensible, intellectual, and moral capacity for pain and pleasure, good and evil, in those who are the objects of right or wrong action from their fellows. It is further extended and varied by innumerable diversities in the conditions of moral influence, and by the differences and contrasts of spiritual condition, which experience reveals in the members of the human family, and of which dim revelations are given us, extending beyond our own limited experience to higher or more distant worlds.

The doctrine that Morals are eternal and immutable may be viewed in contrast to two main errors or falsehoods, which have widely prevailed.

And first, the truths of morality are not to be looked upon as if they were "by will, and not by nature," or arbitrary results from some appointment of the Supreme Creator. This doctrine has been held not only by philosophers but divines, who have strangely supposed that they honoured the Almighty Power of God by advancing a doctrine which makes the Divine goodness a fiction without meaning. The view has been maintained in two different forms. The first of these assigns moral distinctions in the law of right and wrong to a direct, positive appointment of God. He might have enjoined hatred instead of love, selfishness instead of kindness, falsehood and treachery instead of truth, and on the principle of these writers the obligation would equally have followed. It would have been a duty to follow in the wake of the sovereign command; so that, in their hypothetical world, to call evil good and good evil, to put darkness for light and light for darkness, instead of incurring a solemn curse and woe, would have secured and deserved a blessing.

But the same principle has been adopted and maintained, indirectly and circuitously, by some who strongly denounce it in its bare and naked form. They refer all right and wrong, for their immediate source, to certain emotions of approval or disapproval, which rise spontaneously in the human heart. But they refer this constitution itself to the category of positive ordinances of the Supreme Creator. In this way the dependence of Morals on will, and not nature, the doctrine which Cudworth condemns, after being rejected in its simpler and more naked aspect, reappears under a thin disguise.

This great oversight has a signal illustration in the ethical lectures of Dr Brown. He condemns strongly the view of some theologians, who would resolve right and wrong into the one idea of obedience or disobedience to the commands of God. If this view were just, he argues with truth, "what excellence beyond mere power could we discover in that Divine Being whom we adore as the supreme goodness? If the virtue depend exclusively on obedience to the command, He must have been equally adorable, though nature had exhibited only appearances of unceasing malevolence in its Author, and every command He had given to His creatures had been only to add new voluntary miseries to the physical miseries which already surrounded them." But in other passages he looks upon the moral feelings themselves as nothing more than an arbitrary and reversible constitution of the human heart by the Divine Creator, and writes as follows:

"Our minds have been so constituted as to have these emotions, and our inquiry leads us to the provident goodness of Him by whom we were made. God, the author of all our enjoyments, has willed us to be moral beings, for He could not will us to be happy in the noblest sense of the term without rendering us capable of practising and admiring virtue." "Virtue is a felt relation to certain emotions, NOTHING MORE. We speak always of what our mind is formed to admire or hate, not of what it might have been formed to estimate differently. The supposed immutability has regard only to the existing constitution of things under that Divine Being, who has formed our social nature what it is; and who, in thus forming it, may be considered as marking His own approbation of that virtue which we love. Such is the moderate sense of the immutability of virtue for which alone we can contend."

This indirect denial of the essential character of moral distinctions seems to me now, as it did forty years ago, no less mischievous than when the same principle is advanced in its open form. "That which defines cannot, from its very nature, become a property; and thus we cannot justly speak of the Divine Will as good, so long as we regard it as constituting of itself the final standard of right and wrong. But it is plain that we are brought equally to such a conclusion, whether we view it as directly propounding itself for such a standard, or as implanting feelings arbitrarily, to which it afterward refers. The path leads by a longer circuit, but the issue is the same.

Indeed, when searched narrowly, the difference is only that a sense of deception, painful to entertain for a moment, aggravates the evil of the former view. The contrast is the same as between the sway of an arbitrary sovereign, who should claim obedience on the sole ground of his power; and one who should claim further the praise of his subjects for governing with the sanction of a senate of his own appointing, blindly and tamely subservient to his will. Yet surely this would be a faithful emblem of the Divine government, could we conceive the moral emotions to be of mere arbitrary appointment, and liable to be reversed at will, so that what we now condemn as evil we might in such a case have been brought to praise as good. From such a view the mind recoils even more than from the former, and takes refuge in the expostulation of the Psalmist: 'Shall the throne of iniquity have fellowship with thee, which frameth mischief by a law?'"

Again, Morals are eternal and immutable in this further sense, that they do not depend for their truth and certainty on the caprice, the ignorance, or the moral diseases, of those who are subject to their authority.

Pleasures are of many kinds. They may either be pure and healthy, or vicious and diseased. And hence, if moral duty depends on a mere summation of pleasures, and an attempted calculation of their total amount, irrespective of any higher standard, it must be as mutable as those pleasures themselves, which form its component elements. No chain can be stronger than its weakest link. In the view of pure utilitarianism, when the doctrine abides in its native simplicity, and is neither infected nor improved by an attempt to ally it with Stoic or Christian elements, moral right must be as mutable as the capricious likings and dislikings of the most fretful, the most childish, or the most vicious among those who are included in the wide universe of moral agents. It may be inferred logically, from the principle thus laid down, that it is as much one part of moral duty to gratify the lusts of the impure, or the malice of the devilish, as to please the pure and the benevolent, and win the approval of the best and wisest of mankind.

Still further, moral truth is eternal and immutable, because it depends very partially, and in only a limited sense, on those variable elements, which enter into all the experience of human life, and diversify the great laws of duty in their outward application. Truths, however fixed and sure in themselves, contract a seeming variableness when they are wedded to those facts, derived from experience and testimony alone, which do vary continually. The principle extends to every department of applied science. The truths of pure arithmetic are fixed and irreversible, but the data of statistical science are dependent on the accuracy of human observation and memory, or the veracity of human testimony. And the same principle must clearly apply to the higher subject of ethics. There is and must be large room for the exercise of a wise expediency, based on actual experience, when the general doctrines of moral science are brought to bear, in detail, on the actual wants and business of life. The courses of the stars in the sky are fixed and stedfast in their own nature. But when their light shines on the troubled ocean, or even on the calm and peaceful lake that sleeps amidst the hills, the image fluctuates and trembles with every breath of wind that creates a passing ripple or wavelet on the face of the waters.

The treatise of Cudworth, however encumbered with scholastic forms of thought, which detract in some measure from its value, contains some important truths, vital to the integrity of moral science, and no less seasonable at the present hour than when the learned writer first gave them utterance.

And first he maintains earnestly, and with real force of reasoning, that moral distinctions are no creation or appointment of arbitrary power, but are founded essentially on the nature of responsible agents, and of all created being. Descartes had laid down the opposite principle, that the Divine Omnipotence implied the dependence of all things, the truths of geometry, and the distinctions of right and wrong, on some fiat of almighty power, and that "no good or evil can be conceived, of which any idea was in the Divine intellect, before the will of God determined to make it such as it is." On this view Cudworth observes as follows:

"If the natures and essences of all things depend upon a will of God that is essentially arbitrary, there can be no such thing as science or demonstration; nor can the truth of any mathematical or metaphysical proposition be known otherwise than by some revelation of the will of God concerning it, and a certain fanatic faith and persuasion thereupon, that God would have such a thing to be true or false for such a time, or so long. And so nothing would be true or false naturally, but only positively, all truth and science being arbitrarious things. Neither would there be any moral certainty in the knowledge of God himself, since it must wholly depend on the mutability of a will in Him essentially indifferent and undetermined."

"As for the argument that, unless the essences of things, and all truths and falsehoods, depend on the arbitrary will of God, there would be something independent of God, if it be well considered it will prove a mere bugbear. For no other genuine consequence is deducible from this assertion, but that there is an eternal and immutable wisdom in the mind of God, and thence participated by created beings, independent of the will of God. Now the wisdom of God is as much God as the will of God; and whether of these two is best conceived to depend on the other, will be determined from their nature. For wisdom hath in itself the nature of a rule and measure, as a most determinate and inflexible thing; but will being, as considered in itself, indefinite and indeterminate, has the nature of a thing regulable and measurable. Wherefore it is the perfection of will, as such, to be guided by wisdom and truth. But to make wisdom, knowledge, and truth to be arbitrarily determined by will, and regulated by such a plumbean and flexible rule, is quite to destroy the nature of it. Now all the knowledge and wisdom that is in creatures, whether angels or men, is nothing else but a participation of that one eternal, immutable, uncreated wisdom of God, or like so many reflections of one face made in several glasses, whereof some are clearer, some more obscure, some standing nearer, others farther off."

"Moreover, it was the opinion of the wisest of the philosophers that there is in the scale of being a nature of goodness, superior to wisdom, which therefore measures and determines the wisdom of God, as His wisdom measures and determines His will. Wherefore, though some make a contracted idea of God, consisting of nothing else but will and power, yet His nature is better expressed by others in the mystical representation of an infinite circle, whose inmost centre is simple goodness, the rays and area, all-comprehending and immutable wisdom, the exterior periphery or circumference, omnipotent will or activity, by which everything without God is brought forth into existence. Wherefore the will and power of God have no command inwardly on the wisdom and knowledge of God, or on the moral disposition of His nature, which is essential goodness; but the sphere of its activity is without God, where it hath absolute command on the existences of things; and is always free, though not always indifferent, since it is its highest perfection to be determined by infinite wisdom and infinite goodness."

His remarks on the hypothesis of some writers, agreeing nearly with that of Protagoras, that nothing can be known as absolutely true, but only on the assumption that our faculties are rightly made, have a striking application to some later speculations, which deny that any knowledge is attainable of "things in themselves."

"If we cannot be certain of the truth of any thing, but only *ex hypothesi*, that our faculties are rightly made, of which none can have any certain assurance but only He that made them, then all created minds whatsoever must be condemned to eternal doubt. Neither ought they ever to assent to any thing as certainly true, since all their truth and knowledge is but relative to their faculties, arbitrarily made, that may possibly be false, and their clearest apprehensions nothing but perpetual delusions. Wherefore, according to this doctrine, we have no absolute certainty of the first principles of all our knowledge, as that equals added to equals produce equals, or that every number is either odd or even. We cannot be sure of any mathematical or metaphysical truth, or of the existence of God or of ourselves."

"For whereas some would prove the truth of their faculties from hence, because there is a God whose nature is such that He cannot deceive, this is nothing but a circle, and makes no progress. For all the certainty they have of the existence of God and of His nature depends upon the arbitrary make of their faculties; which, for aught they know, may be false. Nay, according to this doctrine, no man can certainly know that there is any absolute truth in the world at all. And this is very strange to assert, that God cannot make a creature which shall be able certainly to know either the existence of God, or of himself, or whether there be any absolute truth or no."

"The ultimate resolution of theoretical truth, and the only criterion of it, is in the clearness of the apprehensions themselves, and not in any supposed blind and unaccountable make of faculties. So that the certainty of clear apprehensions is not to be derived from the contingent truth of faculties, but the goodness of faculties to be tried by the clearness and distinctness of apprehension. For to suppose that faculties may be so made, as to create apprehensions of things that are not, is much like that opinion of some, that all the new celestial phenomena, as of the Jovian planets, and the mountains in the moon, are no real things; but that the clear, diaphanous crystal of the telescopes may be so cut, ground, and polished, as to make all those clearly to appear to sense, when there is no such thing."

"It is a fond imagination to suppose that it is derogatory to the glory of God, to bestow or impart any such gift upon His creatures as knowledge is, which hath an intrinsic evidence in itself... It cannot be denied that men are often deceived, and think they clearly comprehend what they do not. But it does not follow, that they can never be certain that they do clearly comprehend anything; which is just as if we should argue, because in our dreams we think we have clear sensations, we cannot therefore ever be sure, when we are awake, that we see things as they really are."

This doctrine of Cudworth, when rightly explained, and freed from exaggerations which disguise its true sense, that moral truths, or the contrasts of good and evil, of right and wrong, of duty and transgression, are fixed, eternal, and immutable in their own nature, and no mere product of animal instinct or prudential arithmetic, no arbitrary creation of an absolute will, lies at the foundation of all Ethical Science. Whenever it is set aside, Moral Philosophy expires, and nothing but a lifeless corpse or shadowy phantom is left in its stead. Wherever the attempt is made, and gains a wide acceptance, to replace it by truths of a lower kind, drawn from the discoveries of physiology, or the calculations of worldly prudence, the fountains of thought are poisoned, and a moral palsy must quickly seize on all the springs of national and social life. The warning of Him, who is the true Light of the world, applies to every variety of moral speculation, which impinges on the firm, granite-like foundations that underlie the superficial varieties of ethical teaching:—"If the light that is in thee be darkness, how great is the darkness!"

It is a great evil to neglect the monitions of conscience, to silence its calm and secret whispers through the power of vice and selfish passion, and to give the reins, in practice, to the lowest instincts of the human heart. But the evil is far greater, and still more deeply to be deplored, when selfishness is turned into an ethical theory; or when uncertain guesses, in physiology, on the resemblances between man and other animals, are made the excuse for denying conscience altogether, and replacing it by the instinctive craving of men or beasts not to be shut out from the company of their fellows, or by the fear of slaves who dread the lash, and shrink from the penalties of human law through dislike of suffering alone. The logical defect in all such theories is no less striking than the moral degradation to which they inevitably lead. Their advocates need to pass over in silence the very point on which the whole science of morals really depends. The human animal has acquired, in some way or other, a liking for society. But how can the mere indulgence of a social instinct awaken any sense of moral approval? There are many cases in which a refusal to indulge it, and an abstinence from social intercourse in forms diseased or excessive, may be one test and sign of real virtue.

> For Wisdom's self
> Oft seeks to sweet retired solitude,
> Where, with her best nurse Contemplation,
> She plumes her feathers, and lets grow her wings,
> That in the various bustle of resort
> Were all too ruffled, and too much impaired.

Again, the stage may easily be reached of dreading the punishments of social law, and striving to escape from them either by obedience or by concealment. But how is this instinctive dread, too often the parent of fraud, falsehood, and cunning, to rise into the dignity of heroic virtue? Only when, in the voices of human law, the soul catches dim echoes of some far-off music from a higher sphere. And, supposing that a perfect calculation of the results of every action, both pleasant and painful, could be made, how can this sum in arithmetic be raised from its own to a higher level, and be made to infer a moral conclusion, a lesson of right and wrong? Why ought we to prefer the integral to its own differential, the collective series to its first and nearest term? How am I to know that it is right, and a moral duty, to trust my calculation, and stifle the instinctive desire for a present and immediate enjoyment, when the rightness and excellence of pleasure is the basis and postulate of the whole calculation? We cannot rise, without the gift of reason, from mere animal instinct to intelligent prudence, and foresight of the distant future. And without the gift of conscience, or some power of reason to discern a law of good and right binding on the soul by its own authority, we can never mount from the lower level of mere prudence to the sense of duty and moral obligation. The great truths which form the objects of Moral Philosophy are no mere gas-lights of earth. They are stars which shine down upon us from the upper firmament. Their light may too often be clouded and obscured by the mists of earth, and lost for a time from our view. But let the mists be dispersed, and they shine out once more, pure and bright as in the first infancy of the world. And when we follow their sacred guidance, they lead our thoughts upward from this land of strife and shadow where we have often to walk in darkness, to a region of light, purity, and peace, the ante-chamber of His palace who sits enthroned in the beauty of holiness without stain, and goodness without measure, above the water-floods for evermore.

LECTURE XII.

THE NATURE AND OFFICE OF CONSCIENCE.

THE doctrine of Conscience forms the subjective element of Moral Science. It holds a middle place between that aspect of the truth, which dwells on moral laws or principles as fixed in their own nature, immutable and eternal, and that doctrine of consequences, which traces out the connection between the moral character of actions, and results which either may be reasonably expected to follow, or do really follow. The first, when separated from the others, tends to a dreamy, unpractical idealism, the second to a self-satisfied, capricious individualism, the third to a cold, selfish, calculating, mechanical expediency. It is only when the relation of the three principles to each other is apprehended in its real harmony and proportion, that ethical speculation can be freed from the risk of serious evils.

Some of the main questions which may be started respecting the place of conscience in a moral theory are these. First, what is its nature and true meaning? Next, what are the extent and proper limits of its authority? Thirdly, is it capable or incapable of right education? Lastly, if capable of it, on what must its sound and healthy training depend, and what is the external standard which it requires for its guidance?

There are four doctrines, distinguishable from each other, which have been held respecting the nature of conscience. The first views it as an instinct of the animal nature, developed and transformed; the second, as a separate sense or faculty, analogous to the senses of the body, a special and arbitrary endowment of the human mind; the third, as a power of prudential calculation, linking itself, by association of ideas, with strong emotions of hope and fear; the fourth, as the reason itself, exercised on truths of a special kind, or the moral relations between intelligent agents. On this view the distinction is not so much in the faculty itself, as in the character of those truths with which the mind is occupied, whenever we speak of the working of conscience, and the moral emotions of the heart.

The first view of conscience may be dismissed in few words. It has been thus expounded in a recent work. "Man cannot avoid looking backward, and comparing the impressions of past events and actions. He also looks continually forward. Hence, after some temporary desire or passion has mastered his social instincts, he will reflect and compare the now weakened impression of such past impulses with the ever present social instinct, and he will then feel that sense of dissatisfaction which all unsatisfied instincts leave behind them. Consequently he resolves to act differently for the future, and this is conscience. Any instinct which is permanently stronger or more enduring than another gives rise to a feeling, which we express by saying that it ought to be obeyed."

It is strange how such an explanation of conscience or moral emotion can have found acceptance with any thoughtful mind. It assumes the existence in man of those powers of meditation on the remote past and distant future, which distinguish him from mere animals, and constitute him a reasonable being, susceptible of moral laws; and it then proceeds to account for moral emotions, apart from any higher truth, or any relation to these specially human faculties, by the effect of animal instinct alone. Does then every strong impulse, often yielded to, create the sense of a moral obligation? Is it not the voice at once of a frequent experience, and of the deepest philosophy, which is expressed in the words of the Apostle, "The good that I would I do not, and the evil I would not, that I do"? The yielding to a social instinct, which has been violently thwarted, is an impulse of the animal life, but no act of duty; and, standing alone, has no moral character whatever. It needs, in the phrase of Professor Grote, to be "moralized," by the inquiry,—"Is it right or wrong for me to satisfy this instinct? Is it one which ought to be followed, or resisted and controlled?" Only after such a decision does the act assume a really moral character. Feelings of malevolence, in our actual world, are often very strong and very enduring. The Indian will mark his victim for death, and pursue him for years with undeviating aim and unrelenting malice. Does this warrant the conclusion that the instinct is very strong, and therefore ought to be obeyed? Strength and Force, in the old Greek legend, were the servants of Zeus, by whom Prometheus was bound, but the heathen poet never dreamt of exalting them into the parents of all virtue. He looked on them, rather, as the most fitting impersonations of blind, servile obedience to arbitrary and despotic power. It is neither the strength of an instinct, nor its frequent recurrence, which the human mind, unless demoralized by vice or false philosophy, will dignify with a righteous character, and pronounce that it "ought to be obeyed." That strain is of a higher mood. That voice speaks of a higher element in man's being than animal instinct, or the craving for society alone. It is a response and echo, within the soul, to a voice that speaks to it from above. It bears witness to the great truth that man is not merely a magnificent specimen of zoology, but a being only a little lower than the angels, created at first in the image of the invisible and eternal God.

The opinion, which looks upon the conscience as a moral sense, resembling those of the body, or a distinct and peculiar faculty of the mind, deserves a fuller notice. This is the view of Lord Shaftesbury in his *Inquiry concerning Virtue*, and in the *Characteristics*, as well as of Dr Hutcheson, and in substance it seems to be shared by Butler also. "The mind," says Lord S., "observes not only things, but actions and affections. The mind which is thus spectator and auditor of other minds cannot be without its eye and ear, so as to discern proportion, distinguish sound, and scan each sentiment or thought which comes before it." "The true spring of Virtue," observes Hutcheson, "is some determination of our nature to study the good of others, or some instinct which influences us to the love of others, as the moral sense determines our approval." The phrase is based on a true analogy. But, as employed by the writers who introduced it and made it widely current, the truth it implied was mingled with a serious defect. A concrete, positive, and arbitrary character was thus assigned to all our moral emotions and decisions, as if they depended on our being created, through an act of Divine sovereignty, with a sense or faculty of this peculiar kind.

Dr Brown condemns in words the theory of the "Moral

Sense," but adopts substantially the very same view. He writes as follows:

"The great error of Dr Hutcheson and others, who treat of the susceptibility of moral emotion under the name of the moral sense, appears to me to consist in their belief of certain moral qualities in actions, which excite in us what they consider as ideas of those qualities, in the same manner as external things give us, not merely pain or pleasure, but notions or ideas of hardness, form, and colour. It is on this account the great champion of this doctrine professes to regard the moral principle as a sense; from its agreement with what he conceives to be the accurate definition of a sense—'a determination of the mind to receive any idea from the presence of an object, which occurs to us independent of our will.' What he terms an idea, in this case, is nothing more than an emotion, considered in its relation to the action which has excited it. A certain action is considered by us, a certain emotion arises. There is no idea but of the agent himself, and of the circumstances in which he was placed, and the physical changes produced by him, and our ideas of these we owe to other sources. To the moral principle, the only principle of which Hutcheson could mean to speak as a moral sense, we owe the emotion itself, and nothing but the emotion."

Here, as in several other instances, Dr Brown, while professing to detect and remove an error, retains it, and makes it greater than before. "Susceptibility of moral emotion," when compared with "the moral sense," is a more cumbrous and circuitous, and certainly not a more distinct or intelligible phrase. We might call the power of eyesight "a susceptibility of physical emotion from visible objects," but in so doing we should neither improve our style, nor add to the clearness of our conception. The doctrine of a moral sense has its strong and its weak side. Its strength consists in the doctrine that there are moral truths or relations without us, in moral agents and their conduct, which have a certain analogy to the relations of outward objects to each other, perceived by the senses; that these relations are equally true with the other, and that we have a sense or faculty whereby they are perceived. Its weakness consists in the notion that the conscience is a special organ or faculty, like the eye or ear of the body, constructed or implanted for this especial purpose, as the eye for noting form and colour, and the ear for sounds, and in seeming to press this partial analogy so far as to overlook an important contrast. For certainly our perception of rightness or wrongness in actions, and of virtuous or vicious character, is not exactly of the same kind as our sensation that honey is sweet, that fire burns, and ice is cold.

Now Dr Brown seems to oppose or abandon that aspect of Hutcheson's and Shaftesbury's view, which forms its strength, and adopts and amplifies its weakness. The term, sense, may suggest the notion either of truths or facts without us, which we are able to perceive, or of some present, transient sensation and feeling alone. Dr Brown rests his ethical theory on the latter view of the subject. We have certain emotions, however they may arise. We have strong feelings of liking for some actions, and disapproval of others. So far we can go, but no further. Vice, virtue, merit, demerit, obligation, duty, eternal truth, Divine law, are all pronounced to be one and the same fact variously disguised, a susceptibility of the emotions of moral approbation and disapprobation. By this view the subject is said to be freed from much

THE NATURE AND OFFICE OF CONSCIENCE. 247

superfluous argument. But the simplicity is of the same kind, as if we strove to replace all the reasonings and conclusions of geometry, and the wide range of geographical and astronomical discovery, by reducing them all to floating, transient impressions of daylight and colour produced in individual minds. We must first see, before we can observe and discover. We must have a sensation of sound through the ear, before we can arrive at a science of acoustics, or drink in the rich and complex varieties of human speech and musical harmony. There could be no moral science if we were not capable of seeing in our actions, or those of others, those moral features which awaken feelings of shame and indignation, or else of approval, sympathy, and praise. But the very first step of real progress consists in freeing our thoughts from dependence on the fact of the present sensation, and in rising to the apprehension of some abiding truth, which does not cease to exist when we cease to observe it, or perversely turn away from it, but abides unchanged amidst the confusion and uncertainty of our human experience. The analogy of the moral perceptions with those of the eye and the ear, though it needs to be carefully limited, is real and important. But in either case the path of true wisdom is to rise from the subjective sensation to the perception of abiding realities; and then to dwell upon these in quiet thought, till they are freed more and more from transient and accidental circumstances, and stand revealed to us in the calm grandeur of eternal truth.

Others, again, see in conscience only a faculty of prudential calculation, which has to deal with a large problem of the consequences of actions, and learns by habit to associate particular lines of conduct with the fear of punishment and the hope of praise and reward. But while such an estimate of probable future results may supply the materials to guide our actions, as well as our feelings of moral desire and approval, it must still be plain to those who reflect on the course of their own thoughts, or who study carefully the reasonings of utilitarian theorists, that the moral element in all these cases has to be supplied from some other source, and cannot be created by a process of calculation alone. The duty of seeking the greatest good, whether of ourselves or the whole family of mankind, must first be assumed, before we can enter on the addition and subtraction of pains and pleasures, and propose to ourselves, as an aim and object of desire, "the greatest happiness of the greatest number." Once assume benevolence, on the widest scale, to be an imperative duty, a voice of heaven, in some way or other, to the heart of man, and then the laborious attempt to decide what on the whole is best to be done, and ought to be done, may be entered upon with real earnestness, however imperfect the success may be. But until this principle has been assumed, we may have, instead of world-wide benevolence, the mere self-conceit of cold-hearted theorists, who deceive themselves into the notion that they are moral reformers, and are bestowing great benefits on mankind by their fancied skill in a new arithmetic, when they are really indulging the pride of a false philosophy, and undermining the foundations of all genuine morality.

The fourth and last view of conscience is that which looks upon it as no separable faculty, but simply as a name for reason, or the mind itself, when it is occupied with truths of one especial kind. Its supremacy will thus be no arbitrary gift, but result directly from the perceived supremacy of those laws of right and wrong and moral obligation with which it has to deal. It is well defined by Dr Calderwood: "Conscience is that power of mind, by which moral law is discovered to each individual for the guidance of his conduct. It is the reason, as that which discovers to us absolute moral truth, having the authority of sovereign moral law. It is an essential requisite for the direction of an intelligent agent, gifted with free will, and affords the basis for moral obligation and responsibility in human life. In discovering truth, it is seen to be a cognitive or intellectual power. Either it does not discover truth, or if it does, it is not a form of feeling, or combination of feelings, emotions, or desires. Truth is that which we see, and implies seeing power. Moral law is that which we can know, and implies knowing power. The authority of conscience is an abbreviated form for expressing the authority which is common to all the laws of morality. In affirming its authority over the other powers of the mind, we merely indicate that moral law is authoritative for the regulation of all the other motives and restraining forces which operate within our human nature."

The same view of the nature of conscience has been already affirmed by me in another work.

"It may be allowable, as a figure of rhetoric, to speak of conscience as a judge, which holds its court within the soul, and pronounces its judgment on all the lower faculties. But such metaphors, when constantly used, are liable to create a serious delusion. Conscience is simply the mind itself, exercising its judgment on the moral relations of right and wrong, on its own actions and the actions of others. Its supremacy over other faculties is merely a varied expression for the truth, that the relations the mind contemplates, when its acts receive this name, are in their own nature of binding authority, and claim allegiance and submission. In its other actings the mind contemplates things equal or inferior to itself, or superior beings, irrespective of any claim to actual dominion and supremacy. But the laws of moral duty are royal laws in their own nature, and speak with the voice of a king; and the judgments of the mind, in which it recognizes them, must partake of the same character. Thus the supremacy of conscience depends entirely on the distinctive nature of moral truth; but its defects, weakness, and error, are due to the mind itself, and are one form of its moral guilt and infirmity. Its dictates are binding so far as they are genuine reflections of eternal truth, or of real moral relations perceived by the soul. But the mistakes of conscience have no more real authority than any other kind of error. They have this peculiar nature, that they make sin inevitable. In obeying them, man sins against the laws of God; in disobeying them, against his own convictions of duty, and the internal harmony of his moral being."

On a just view of the nature of conscience must depend the answer to the question, what are the extent and proper limits of its authority? Is it supreme in any sense? And if so, is it absolutely or only relatively supreme?

"The high honour" has been claimed for Bp. Butler "of establishing the supremacy of conscience." And he certainly lays it down with great clearness and force in these well-known words.

"That principle by which we survey and either approve or disapprove our own heart, temper and actions, is not only to be considered as what is in its turn to have some influence, which may be of every passion, and the lowest appetites; but likewise as being superior, as from its very nature manifestly claiming superiority over all the others; insomuch that you cannot form a notion of this faculty, conscience, without taking in judgment, direction, superintendency. This is a constituent part of the idea of the faculty, and to preside and govern, from the very constitution of men, belongs to it. Had it strength as it has right, had it power as it has manifest authority, it would absolutely govern the world." (Serm. II. p. 36.)

This claim, however, of conscience to the governing office, and a rightful authority over all the powers and actions of mind and body, seems too plainly involved in its nature, and too elementary a truth, to be rightly assigned to any modern author as his own discovery. All that Bp. Butler could do was to reassert a familiar truth, involved in the very conception of a moral law, and its apprehension by the mind of man. He has unfolded it in contrast with one special form of immoral reasoning, which would set up a democratic equality of all the desires of the mind, or appetites of the body, simply because they exist. Such a republication of old truths in some varied aspect, to meet fresh phases of error, may be very important in its place. But we honour the moralist at the expense of the great science with which he deals, when we represent its fundamental data to be so obscure as to have been first discovered in modern times. These truths are old as the creation. The supremacy of the moral law, on which that of conscience wholly depends, is revealed in the books of Moses, long ages ago, more clearly and powerfully than in the sermons of Butler, or in the treatises of heathen or Christian moralists of later times.

The authority, it has been well observed, "is not found in the nature of the faculty itself. The faculty is a power of sight, such as makes a perception of self-evident truth possible to man, and contributes nothing to the truth which is perceived. To the truth itself belongs inherent authority, that is, absolute right to command, not force to constrain."

The supremacy, then, of conscience is limited and relative, not absolute. The eye does not create the landscape on which it gazes. Neither does the individual conscience create the moral truths it contemplates, or clothe them with their claim to obedience and honour. The precepts of genuine morality, natural or revealed, are of direct and immediate obligation. They bind, because they exist, and are the voice of God. They can be felt to be binding, and guide the practice, only so far as their authority is accepted, and their true meaning discerned. A personal conviction with regard to our own duty neither adds to their authority, nor creates the obligation to obey; just as an image on the retina does not really intervene between the eye and the landscape, but is only the necessary condition, from the structure of the eye, on which healthy vision depends.

Is the conscience, as a faculty of the mind, capable of education and training? To this question Dr Calderwood, whose definition I have quoted, gives a negative answer. In this he dissents, as he owns, from many authors with whom he substantially agrees in other things, as Reid, Dugald Stewart, and Dr Whewell, but quotes Kant on the other side. Conscience, he says, from its very nature, cannot be educated. Education in the sense of training is impossible. As well propose to teach the eye how and what to see, and the ear how and what to hear, as teach reason how to perceive the self-evident, and what truths are of this nature. All this has been provided for in the human constitution. And he then quotes with approval the strange dictum of Kant, that "an erring conscience is a chimera." (Cald. *Handbook*, p. 81.)

The assertion of Dr Whewell, here condemned, is perfectly true, and the opposite statement an error surprising in an author of such clear good sense and ability. No doubt, if we begin by confining the name of conscience to the sound and healthy conclusions of the mind on moral questions, to its clear and distinct vision of spiritual truth, or its perfect insight into the course of practical duty, then all need of training is excluded by the definition. But the definition itself will then degenerate into a barren truism. The mind needs no training to judge rightly, wherever right judgment is already attained. It needs no increase of light, where it already sees clearly. But this is neither the popular, nor yet the scientific meaning of the word. We certainly do not mean by conscientious convictions those which are infallibly right, but simply those which are honestly entertained. A person follows his conscience when he does what he sincerely thinks to be his duty, though he may have mistaken his duty, and acted on a wrong judgment. To the authority of Kant, whose paradoxes on this subject are many, we may oppose the far higher authority of the inspired Apostle, himself the most intensely conscientious of men,—"To them that are defiled and unbelieving is nothing pure, but even their own heart and conscience are defiled."

The illustration to which the appeal is made proves the exact converse. The eye does need to be trained, as in the case of the sailor, to discern distant objects, or as with the artist and his disciples, to catch the features of some magnificent landscape; and the ear is capable of and needs much training, to receive the more delicate distinctions and harmonies of sound. Neither organ would admit of training at all, if the sense were wholly wanting, as in the deaf and the blind. But those who are dim of sight and dull of hearing may do much, by artificial helps, to remedy, and in some cases to remove, their natural infirmity, while careful training how and what to observe is an apprenticeship required in almost every field of scientific inquiry.

I venture, then, with all confidence, to reverse the statement I have quoted above, which rests only on a deceptive limitation in the use of a familiar term. Conscience is a faculty which, like the eye and the ear, does require education, but for more urgent reasons, and to still nobler ends. The Apostle speaks of those who "by reason of use have their senses exercised to discern good and evil." Eyes, which are suddenly brought out of thick darkness, cannot at once bear the full light of clear noonday. They must first be shielded from excess of light, and gradually taught to use their newly acquired possession. For sight, though we speak of it under the name of intuition, to denote what is most immediate and spontaneous in the action of the mind, is really the integral of an immense number of experiences gradually acquired. We see the book, the room, the landscape before us. The perception is swift, like the motion of that light which is the medium of vision, with its billions of vibrations in a single second. But it is really the summation of countless elements of knowledge, in which form, colour, perspective, distance, natural properties, the indications of sight, sound, touch, smell, are all fused and blended into one harmonious whole. And in each direction the sense is capable, by diligent use, of further development. The geologist sees much in every stone, or layer of earth, or mountain escarpment, the naturalist in every plant, or beast, or bird, which does not meet the eye of the common observer. Rich and various associations of every kind gather around the objects, on which the eye rests with fond and deepening interest. Our minds, taught through the senses, lend to nature its wedding garment or its shroud, while they perceive the sad and joyful associations with which every scene is peopled by thoughtful and intelligent observers. The senses do not create, as the bee does not create the honey which it sucks from innumerable flowers. But they lay in growing stores by constant diligence of observation.

Conscience is the eye of the soul. If it be single and upright, the whole moral being will be filled with light. By wakeful vigilance it makes continual accessions to the soul's perception of truth. The landscape is wide and various. Every part of it yields new treasures to the observer, in proportion as he gazes more thoughtfully on the objects submitted to his view. In its first and early stage of activity, when it is like a child still ignorant of the outward world, or a prisoner just escaped from some dark cell, it may scarcely be able to see at all, and may need the help of others more experienced, or higher in moral attainment, before it can detect a thousand features of the objects which lie plainly before it. But the truths contemplated are real, fixed, and abiding truths. The power of the mind, by which it discerns them, is a noble gift of God, a faculty which may be dulled and clouded by vice and sensuality, but is never wholly and hopelessly withdrawn. And the constant training of that faculty, by use and exercise, to fulfil the solemn duties confided to it, is not only possible, but highly expedient, and itself a primary moral duty of ceaseless obligation.

Whoever has no longing for clearer insight into the laws of righteousness, and does not strive at all for their fuller apprehension and practical attainment, must be sinking into a moral lethargy, the dangerous precursor of disease and death.

But if the conscience is capable of education, and requires to be exercised and trained, how is this great end to be secured? Where shall we find a standard, to which the appeal can be made? How are we to detect deviations from this standard, and raise the conscience, when subject to illusion, distortion, or paralysis, to a higher level? This is clearly a question of the greatest practical importance. For though Kant may have uttered the paradox, that "an erring conscience is a chimera," unless the laws of language deceive us on the one side, or the lessons of universal experience on the other, an erring conscience has been, in every age of the world, the most fruitful parent of strife, passion, warfare, and every variety of human folly and crime.

That school of ethical teachers, who would lay the foundations of morals in some recent discoveries or guesses of modern anatomy and physiology, call loudly on the independent or ideal moralist to produce a standard, which on their principles it is impossible to find. Where, they ask, is this rule to be found? "What is the standard conscience? That must be got at, or morality is not a subject to be reasoned or written about...We have standards of length, of weight, of measure, which, even though embodied in material objects, can hardly be said to have the independence contended for...When we are called upon to set our watches by Greenwich time, it is not a standard beyond humanity. The collective body of astronomers have agreed on a mode of reckoning, founded on the still more general recognition of the solar day as the principal unit. At Greenwich observatory observations are made, which determine the standard of this country; and the population, in accepting that standard, know or may know that they are following the Astronomer Royal, and the body of astronomers generally."

To this challenge Professor Maurice gives a partial and indirect reply by an appeal to two fictitious characters, Tennyson's *Northern Farmer*, haunted by strange death-bed scruples, and Tito in *Romola*. He infers from these excursions of fancy, in its truest and most faithful pictures of human life, that conscience is a fact and not a fiction, however hard its proper standard may be to find. But I think we may go considerably further and find, in those very illustrations from science to which the objector appeals, some real help to remove the phantoms of hopeless doubt and uncertainty which are raised by the necromancy of false theories, to crowd around our path, and hinder our clear perception of moral truths.

And first, how is it that, even in the business of daily life, we need standards of length, weight, and measure? Surely this is a sign, in itself, that men in all their social activity need some common, objective truth, to which all may appeal. They must have measures free from the caprice, the ignorance, and the perverted selfishness, of the individual buyer or seller, or else the wheels of commerce would stick fast in the mire, and the business of life come almost to an end. We thus escape from the vague guess-work of barbarous times, in which every man's arm or foot was his own standard, and attain something firmer, more definite, and more exact. And the standard is better, not because many conspire to receive it, but for a more solid reason. Patient skill and thought have been employed, to get rid, as far as possible, of the sources of error, to secure the best materials for the rod with which other lengths are compared, to guard against possible flexure, to allow for changes of temperature, and approach nearer and nearer to an ideal of fixity, which science teaches us, after all, we have not perfectly attained, while it no less plainly assures us of the reality and closeness of the approximation.

But these standards, it is objected, are arbitrary in their own nature, however superior to the rules and measures they may have displaced. And the reason is plain. Length, solid capacity, weight and time, have no absolute unit. They are matters, not of pure reason, but of practical observation and experiment. They stretch out continuously between zero and the infinite. Before we can compare them, a unit must be assumed in every case, and the choice is a matter of convenience, custom, and usage alone. The relation of the series of numbers to the dimensions of space, the force of gravity, and the sequence of time, is not fixed but variable. It belongs to the province, not of absolute reason, but of intelligent will. The relations of numbers, in themselves, are fixed and unalterable. We cannot even conceive them to be reversed. So are also the relations of space and figure, of force and time. But the link which connects the first with all the rest does not belong to the category of pure reason, but is borrowed from the domain of positive appointment, and of human choice and will.

Even here, when once the choice has been made, there is a ceaseless effort of science to eliminate error, and attain a standard of ideal perfection. The first aim is to secure that every copy shall agree with the original standard, at least within limits practically insensible in the business of life. This could not be done, if lengths, weights,

THE NATURE AND OFFICE OF CONSCIENCE. 259

measures, intervals, because each involves one arbitrary unit, were in themselves capricious and arbitrary things. The yard, the metre, the pound, the gramme, and in fresh branches of science the units of work and of electric tension, are, if possible, to be everywhere the same; and all secondary measures are tested, corrected, and improved, from time to time, by comparison with the originals of which they bear the name.

But the effort proceeds still further. The primitive measure of length or capacity, in its concrete form, is seen and known to be liable to various changes, which interfere with its use as a perfect standard. The rod may bend, and the flexure, though slight, will cause a real variation of length. It enlarges with heat and contracts with cold. The measure of weight is still more variable. It is affected by every change in the unit of length to which it is referred, and has other weaknesses of its own. It is affected by the buoyancy of the air, which varies with temperature and other changes, and still more by the latitude on the earth's surface. Thus every standard ceases to be an external, physical reality. One arbitrary element must remain from the nature of the objects to be measured, but all beside is scientific and ideal. The rod, by which length is measured, is the unit under ideal conditions of temperature, seldom or never realized in the actual measurement. Or else it is a fixed, ideal ratio to some ideal pendulum, vibrating in an ideal latitude, and at an ideal level, in a given part of the time of the earth's revolution; or is some decimal fraction of an ideal elliptic quadrant of the surface of the earth.

In the standards of time this progressive effort is still more striking and instructive. The loose impressions, which vary with every mood and phase of individual thought, are first replaced by a general reference to that universal standard, the natural day. But the sunrise and sunset vary widely with the times of the year; and the hourglass and clepsydra are invented to divide more exactly the hours and intervals of each day, which were more rudely guessed before from the place of the sun in the sky. Clocks and watches are invented, and these are successively tested and verified by the growing exactness of celestial observations. The earliest and most popular unit of time, the natural day, when completed by the night, or reckoned from noon to noon or midnight to midnight, is found by science to be more stedfast than the other celestial cycles. But, to secure fuller accuracy, one important change has to be made, and the solar is turned for popular use into the mean solar, and for observations into the still more perfect standard, the sidereal day. And even here the striving after a perfect standard does not cease. However stedfast the earth's rotation, it has been lately inferred that, from tidal friction, some slight change may still occur in the course of ages. And an attempt has therefore been made, by ascertaining the rotation of Mars to the fraction of a second, to eliminate, by a comparison of the days of two neighbouring planets, the errors that might possibly arise from accepting either of them, alone, for an invariable measure.

Now all this process, by which standards of length, weight, measure and time, are first constituted, and then verified and improved, tends not to confirm but to disprove the dogma of those whom Plato calls "flowing philosophers," by the latest of whom this appeal is made. These facts lend no countenance to the ignoble theory that distinctions of right and wrong are the creations of human fancy, or the result of artificial associations of thought, due solely to the positive appointment of rules by human lawgivers, and to the dread of social punishment alone. Arbitrary units do not enter into the domain of pure science. In arithmetic, geometry, and mechanics, our reasoning is wholly independent of the yard, the pound, the metre, the gramme, which may be needful in the business of actual life. The conclusions depend either on the numerical series itself, or on proportions of distance, weight, force, and figure only. In all practical science these units are needful, and link the truths of pure reason with the realities of the actual world. Actual weights, lengths, and intervals, can all be conceived to have been different, and are to be learned by observation alone. But when once the transition is made, the whole aim and struggle of science is to restrict, within its narrowest limits, the uncertainty which attends the entrance of this arbitrary element. A separate science has been devised, to eliminate the errors of observation in the most effectual way, and to combine the separate data of the experiments in such intimacy of relation to each other, that the result may approach as nearly as possible to the character of pure and scientific truth.

The astronomers in our observatories do not create either the solar or the sidereal day. The accurate measure of time, which the nation obtains through their labours, is no creation of the scientific observer. He merely gains, by careful experiment, a more exact knowledge of that diurnal rotation of the earth, which was a settled fact of nature before observatories were built, or telescopes and transit instruments were made. The natural standard exists, not fixed in its own nature, but by the present constitution of our system. The task of the man of science is to discern it clearly, and to correct, by means of its evidence, carefully observed, the various errors which attach to looser reckonings of time by mankind at large.

The principle in Ethics is the same, and is only varied in its application. Pure morals, like arithmetic and geometry, involve no arbitrary unit, but occupy a higher level of fixed and absolute truth. The great laws of duty, and the contrast of moral good and evil, are inseparable from the nature and existence of agents endowed with reason and free will, who are capable of mutual injury or blessing. They are firm and sure, like the properties of number and space, and like these, open a wide and immense field for the attainment of certain truth. A round square is not more inconceivable than the union of causeless and malicious hate, or of deceit and treacherous cunning, with the attribute of perfect goodness. But when we pass from the great law of love to the realities of moral action in the world around us, an arbitrary element begins to appear. That law, as divinely revealed, includes a reference to the actual powers and constitution of those who are subject to its authority. Where the strength is greater, and the heart and mind and soul are more largely gifted, the precept of duty, to love with all the mind and strength, must assume a larger compass, and range over a wider field. And when we descend lower, and enter on the complex relationships of life, the positive and variable elements become still more numerous. In every circle of human society, whether more narrow or comprehensive, in the family, the church, and the state, secondary standards, the product of local custom or national legislation, need to be extensively employed, and these may vary more or less from each other. The principle of obedience to laws ordained by a just authority is the same, but the laws may be varied to meet the varying circumstances of national and social life. Yet this is no proof or sign that they are founded on caprice and the selfishness of power alone. We might infer, with equal truth and wisdom, that the labours of our royal astronomers had created the stedfast order of the earth's revolution, and of the celestial movements. A perfect standard, wholly free from arbitrary and positive elements, is only possible in that highest department of morals, where it dwells in the light of heaven, and leaves behind it all the specialties of earthly life in the presence of God's eternal throne. But even where such perfection is not attainable, a ceaseless effort to approach nearer to its attainment is the first instinct of all genuine science. The lazy and the dissolute may abandon the search, and resign themselves, either in wilfulness or despondency, to the loose guesses of capricious ignorance, or the blind impulses of self-will. But true morality, like all genuine science, strives ever to ascend from the fleeting, variable phenomena, where no rule or law is seen, to the perception of a law that abides in the midst of perpetual change, and a standard pure and absolute, on which the secondary copies, useful for the guidance of daily life, depend for all their real worth, by which they must also be tested, when they are found to vary and contradict each other. For even human standards of length or measure are not mere arbitrary inventions. They point silently and secretly to a fixed relation in the nature of things, without which the attempt to provide them for general use, and then to correct their errors, would be impossible. The words of Cudworth include a reply, by anticipation, to the sceptical inference falsely drawn from human reckonings of time.

"A horologe is not mere silver and gold, brass and steel, any way mingled or confounded together. It is such an apt and proportionable disposition of these materials as may harmoniously conspire to make up one equal and uniform motion; which runs parallel, as it were, with the motion of time, and, passing round the horary circle, and measured by it, may also measure out that silent and successive flux, which, like a still and deep river, carries down all things along with it indiscernibly and without any noise; and which in its progressive motion, treads so lightly and softly that it leaves no traces, prints, or footsteps at all behind it."

Thus all human standards of weight and time and measure imply a deeper law or fact of nature on which they depend. And human conscience, in like manner, in all its diversities and partial errors, points upward silently to a law of eternal and unchanging righteousness, whose seat is the bosom of God, and whose voice is the harmony of the world, the music of the celestial spheres.

LECTURE XIII.

THE DOCTRINE OF UTILITY.

A THIRD element, which must enter into all inquiries concerning the foundations of Moral Science, consists in the relation between moral acts themselves and the results or consequences to which they lead. The tree is known and tested by its fruit; and the nature of the fountain, by the streams which flow from it, and the verdure or barrenness of the territory through which they flow. When this element is wholly set aside, as in some schemes of "intuitive morals," ethical teaching will be likely to assume a vague, dreamy, and unreal character, and will awaken natural distrust in practical minds. On the other hand, when this view of the subject is made the one sufficient key to unlock all the secrets of conscience, and the mysteries of human life, all the highest and noblest aspirations of the soul are liable to be lost and buried in a quagmire of calculations impossible to be performed, from which extrication is hopeless. Such is the fundamental defect of all purely utilitarian theories.

Every moral action has a threefold relation, to a precedent law of right and duty, to which the agent owes allegiance,—to a present faculty of the agent himself, whether called conscience, the moral sense, or the practical reason, by which the law is perceived, and becomes applied to his own heart and life,—and to the results in his own later experience, and in the happiness or unhappiness of others, to which right or wrong conduct, a virtuous or a vicious life, is sure to lead. And this triple character of all human action is only the reflection of that higher truth, which sums up the noblest conceptions of Christian Theology:—"For of Him, and through Him, and to Him are all things, to whom be glory for ever. Amen."

The doctrine which makes utility the exclusive test of right and wrong, and decides on the moral character of actions by their supposed or expected consequences alone, may assume very different forms. The first is the theological, in which it borrows, but in borrowing distorts and partially degrades, some great truths of the Christian revelation. God wills the happiness of mankind. He commands us to practise universal beneficence. His will is sanctioned by promises and threatenings, that are to be fulfilled in a future life. Therefore self-love requires us to obey His command, and to practise works of social kindness, in hope of gaining the promised reward. But in applying the principle we are left to our own judgment; and the known tendencies of actions, deduced from experience, are said to be our chief guide.

The second form of the doctrine is the philanthropic or benevolent. All the motives of religious faith are either formally set aside, or silently disappear from view. In their place there is borrowed from the rival doctrine of intuitive morality a vast *à priori* maxim, the supreme, essential obligation, needing no proof, and assumed to be self-evident, of universal philanthropy or benevolence. But this great principle, whether borrowed from Christianity or from philosophic idealism, is no sooner assumed, than it is disguised, concealed, and consubstantiated, under the form and accidents of a complex process of experiment and calculation. The whole business of morals is to calculate results, and work out problems of maxima from imperfect premises; while the one element which alone has a truly ethical character, the deliberate, earnest, conscientious aim to do good to our fellows, and in so doing to please and serve the common Creator and Preserver of mankind, is left habitually out of sight, lest it should embarrass and disturb that process of arithmetic on which the whole science is made to depend.

The third form is the philosophically selfish, or that of Epicurus and his later disciples. It recognizes no religious faith in its scheme of morals, nor any need for motives drawn from the Christian message of a life to come. Neither does it purloin from the Scriptures the second great commandment, and then conceal the precious treasure, as Achan hid the talent of gold in the soil of his tent, burying it in the heart of a system of pleasure-seeking arithmetic, with which it has no natural connection. It lays down the principle, based on certain animal instincts, that the attainment of personal pleasure is the main end and business of life. And then it proceeds to mitigate the harshness, and prune away the grossness, of the naked theory, by insisting on the need of a wise and thoughtful prudence, grounded on the lessons of experience, to free men from the pursuit of vicious indulgence, and to prove the superior gain of temperance, kindness, the restraint of passion, and the cultivation of private friendship. And there can be no doubt that the laws of prudence, when really studied and observed, may form the first steps in an upward progress, from which the mind must, soon or late, gain clear glimpses of higher and holier laws of action than the pursuit of selfish and personal pleasure alone.

The fourth and last form is that of political selfishness. Virtue, on this view, consists in a habit of submission to outward laws, created and sustained by the fear of human punishment. Instead of rising above the love of fame,

That last infirmity of noble mind,

it consists rather in one of the worst infirmities of minds both feeble and ignoble; that is, in the animal fear of physical suffering, engrained and engrafted in the heart by cultivating the habits and instincts of a slave.

The attraction of all these theories lies in the desire to escape from the caprices of individual conscience, and uncertain standards that may pass current in society from time to time, and to obtain for morals some firm and solid basis, which does not vary with the fashions of the day. From the ethics of common sense and instinctive feeling we may either rise towards the stars, or descend to find safer footing on the solid earth below. Thus we are told that "utility sets up an outward standard in the room of an inward, being the substitution of a regard to consequences for a mere unreasoning sentiment or feeling." And again, that "the contest between the morality which appeals to an external standard, and that which grounds itself on internal conviction, is the contest of progressive morality against stationary, of reason and argument against the deification of mere opinion and habit." In reality, it is needful both to rise and to descend; to gaze on the skies above, and also to-tread firmly on the ground beneath our feet. But when philosophy would teach us that, to attain certainty, we must not only tread on the ground, but that our eyes must "be always downward bent," to explore the buried riches of the soil beneath us, we may remind it that the discoveries of genuine science often reverse the too hasty conclusions of the senses, and that fixity is rather to be found in the laws of the starry heavens than in the swiftly moving, ever-changing surface of that earth on which our feet seem to rest so firmly below.

The principle, which forms the basis of the utilitarian doctrine, has been laid down by Bentham, in his *Theory of Legislation*, in these words:—

"A principle is a first idea, which is made the beginning or basis of a system of reasonings. To illustrate it by a sensible image, it is a fixed point to which the first link of a chain is attached. Such a principle must be clearly evident. To illustrate and explain it must secure its acknowledgment. Such are the axioms of mathematics: they cannot be rejected without falling into absurdity.

"The logic of utility consists in setting out, in all the operations of the judgment, from the calculation or comparison of pains and pleasures, and in not allowing the interference of any other idea.

"I am a partizan of the principle of utility, when I measure my approbation or disapprobation of a public or private act by its tendency to produce pleasure or pain; when I employ the words *just, unjust, moral, immoral, good, bad*, simply as collective terms, including the ideas of certain pains or pleasures; it being always understood that I use the words, *pain* and *pleasure*, in their ordinary signification, without inventing any arbitrary definition for the sake of excluding certain pleasures, or denying the existence of certain pains. In this matter we want no refinement, no metaphysics. It is not necessary to consult Plato or Aristotle. Pain and pleasure are what every one feels to be such, the peasant and the prince, the unlearned and the philosopher.

"He who adopts the principle of utility esteems virtue to be a good, only on account of the pleasures which result from it; he regards vice as an evil, only because of the pains which it produces. Moral good is *good* only by its tendency to produce physical good. Moral evil is *evil*, only by its tendency to produce physical evil."

This geocentric theory of morals, however captivating to some minds by its appeal to the senses, and its air of simplicity, will be found, when closely examined, beset with difficulties of the most formidable kind. It does not really secure that certain and solid footing, to which it sacrifices some of the noblest instincts of the soul. Utility in morals, like fire in the domestic economy, is an excellent servant, but a bad master. Its enthronement as supreme is the destruction of the science which it pretends to adorn and improve.

The first defect, its merely external character, meets us in the outset of the theory. It deals only with the outward actions themselves, and professes to dispense, almost entirely, with a reference to the inward motives and feelings of the heart. "There is no point," it has been said, "which utilitarian thinkers have taken more pains to illustrate than the difference between motive and intention. The morality of the act depends entirely on the intention, that is, upon what the agent wills to do. The feeling which makes him will so to do, when it makes no difference in the act, makes none in the morality, though it makes a great difference in our moral estimation of the agent." (Mill's *Util.* p. 27.)

Here, then, we have an attempt to build a scheme of moral action, from which all respect to the motives of the agent is excluded as a superfluous element. But the attempt is deceptive and vain. Moral elements are needful to constitute a moral or immoral act, and these elements are to be found in the motives and feelings of the agent, and in these alone. The contrast of motive and intention is only that between a more immediate object, and one more remote. He who intends to do good to another, and harms him in the attempt, may be guilty of a serious wrong. But the wrong thus inflicted is wholly different, in a moral point of view, from similar acts done in deliberate malice. There may be perhaps a criminal rashness, a blind and erring superstition, or a disgraceful ignorance. The act may thus be worthy of moral blame. The laws of the land may sentence it to a just punishment. They may even, from the act itself, infer a constructive malice. But the moral distinction, wide and deep in itself, still remains. The motive is the soul on which the true nature of every act mainly depends. But if there be grievous negligence in forming the judgment, so as to inflict injury where kindness was meant, this fault is a moral evil of another kind. And when this neglect is wilful and aggravated, the man may be justly held responsible, almost as much as for malicious wrong, for evils which his ignorance and rashness may have caused to the objects of a well-meaning, but misguided, thoughtless, and perverse benevolence.

The doctrine, which makes the moral good or evil of actions depend only on their consequences, is open to an objection at the outset of a very fatal kind. Acts, viewed merely as physical changes, can have only physical consequences. A sword is drawn, a pistol is fired, and death ensues. Set aside all reference to the human will, the motive of the agent, and in either case there is a physical result alone. So much matter is moved, slowly or swiftly, from place to place. Chemical elements enter into new combinations. Light, heat, and sound are generated, swift and sudden motion is given to a ball of solid metal, muscles or nerves are shattered, the conditions of vital action are violated, and there is a lifeless corpse, instead of a living man. But in all this there is no moral consequence or result whatever. Once introduce the idea of will and choice, and all is altered. The agent is no longer a mere automaton in a vast engine-house of nature, full of machines. He has acted rightly or wrongly. He has done good or evil. He has committed an act of lawful self-defence, or of patriotic effort, or has been guilty of a culpable homicide, or even of a deliberate and malicious murder. The motives of the agent, and his relation to the person slain, alone give the act its moral colouring, and render it the fit object of approval or blame.

All actions are thus morally sterile and unproductive, unless moral features, prior to any later result, are assumed and pre-supposed. They can never, when these are wanting, rise above the level of physical changes, morally indifferent. We must recognize in them some working of free choice, of a responsible and intelligent will, of the desire to injure or benefit others, to do good or to do evil. Then the dead corpse receives a new and higher life. They become moral acts, out of which moral consequences without number may be expected to flow. A blow, in itself, is only one insensible element in a vast problem of molecular change and planetary perturbation. But let it be an act of wilful and deliberate insult, and it may lead to some fatal duel, or possibly wrap a whole kingdom in the flames of civil strife, or carry war and destruction to countless families and unborn generations.

Again, the moral tendency of actions presupposes and requires moral laws of right and wrong, which affect not the agent only, but those who are the objects of his activity. The phrase sums up briefly all the effects that follow, when injuries or benefits awaken various emotions of gratitude or ingratitude, of indignation, resentment, or forgiveness, in those who have been affected by them. It would be a vain attempt to trace dynamical consequences, if we first assume a universe of particles wholly inert, in which the forces of gravitation and cohesion and electric affinity have no power. The persons acted upon must have capacities of moral discernment, before moral results, worthy of the name, can follow. It is the perception, whether true or false, of some moral character in the act, by those whom it directly or remotely affects, on which the nature of the consequences will almost entirely depend. And hence, in a world where ignorance, error, and falsehood prevail, the estimate of actions by their immediate and apparent consequences is liable to be most deceptive. The clouds will hide the sun. The drifting meteors project their own vagrancy, in parallax, on the orbs which are stedfast; and virtues which cross the current of popular vice, and awake the opposition and hate of the corrupt and selfish, may be mistaken, like ancient prophets and apostles, for guilty troublers of society, criminals that turn upside down the peace of the world.

There is a further disproof of the doctrine that a calculation of consequences is the only source of morals, from reflection upon the nature of the calculation itself. It is a moral act, and subject to moral laws. The power to foresee consequences, to remember the results of past experience, to compare and analyse the causes of events, and to trace the probable effects of a particular course on the mind and feelings of others, is a high and noble faculty.

Man did not create it for himself, it is the gift of God. If actions themselves need moral rules to be calculated for them, how much more the faculty by which these calculations are to be made! But who can learn, by calculating alone, the methods and principles on which all right calculation must depend? *Quis custodiet ipsos custodes?* Surely the gift is one of which the exercise needs some guidance, since on its right or wrong performance, by the hypothesis, a life of vice or virtue, of wisdom or guilt and folly, must naturally ensue.

The doctrine which assumes that pleasures are to be courted simply because they please, and suffering to be avoided simply because it is painful, turns a mere animal instinct into a fundamental rule of moral arithmetic. On what warrant is this rule assumed? First principles, we are told, must be clear and evident, like the axioms of mathematics. And then it is assumed, in the next paragraph, that pleasures of disease, of vice, and malevolence, are to enter into our calculation side by side with the pleasures of Christian piety or social kindness, and must weigh equally in the scale, if their amount or quantity be the same. But a calculation of results, based on such a confusion of moral opposites, is immoral in its own nature. Instead of founding a system of genuine ethics, it may be said to involve a guilty and fatal apotheosis of vice, disease, and folly. All reckoning of moral consequences is the use of a high and noble faculty of man's being. It is not a lawless process, to be conducted by the capricious decisions of an erring philosophy, when it confounds or denies distinctions on which the foundations of morality depend. It is subject to laws of moral duty. The pleasures to be compared must be tried by a higher standard than of their seeming intensity alone. Factors introduced by human vice and folly must be thrown aside, since they only tend to lower the tone of thought, and to prevent any true solution of a hard problem. For surely it is no less immoral to accept the diseased pleasures of others, their corrupt and malevolent passions, or their gross and sensual practices, for positive elements to guide my actions by an attempt to increase and enlarge them, than to indulge the like pleasures in my own person.

Again, the process of calculation, so far as it is lawful and desirable, is subject to limitations of a moral kind. There are cases in which it may be right and wise to delay immediate action, in order to gain a clearer judgment on the course which is best to pursue. But there are many others, in which prompt and immediate action, even with risk of partial mistake, is better than protracted deliberation. Now if the moral rightness of an action depends entirely on a correct estimate of all its consequences, in all such cases to act rightly must be impossible, because the requisite conditions could never be fulfilled. There is a time to speak, and to keep silence. There is also a time to calculate probable results, and a time when it is needful to act promptly and speedily, and long deliberation would be a fatal error. In the great majority of the practical duties of life, it is far better to use our eyes at once, or to correct their known defects with such glasses as are at hand and in constant use, than to spend months and years in constructing some moral telescope or microscope, which, after all, we may never be able to employ, because the season for action has passed away. When the conscience is well trained, and dwells in an atmosphere of purity and light, it is safer to act on its first and simplest impulses, than to submit them to a process of laborious and intricate calculation, which may serve, too often, for a chemical freezing mixture to all the warmer emotions and more generous instincts of the human heart.

Actions, then, can have no traceable moral results, unless moral laws, affecting the agent himself and the objects of his action, are presupposed. They sink to the lower level of a mere physical change, the proper subject neither of praise nor blame. The calculation, if made, must itself be subject to moral laws, and restrained by moral limitations. It is wrong to calculate and plan for the increase of such pleasures in others as it is criminal to indulge in our own person. It is foolish and wrong to go on calculating possible and remote results, when prompt and speedy action is the plain requirement of duty. But the difficulties of the theory are greater still. The calculation of consequences, when set up for the true foundation of morals, and the test of right conduct, is impossible. It is capricious and uncertain in the elements on which it depends. In the form in which it has usually been proposed, it is degrading and immoral. And while the professed aim is to gain a clear and intelligible basis for morality, by such a process we sink in miry clay, and can never arrive at firm and solid ground.

And first, the calculation, viewed on the side of science, is impossible. It requires the summation of an infinite series. And the series is one of which the laws, as borrowed from experience only, are so immensely complex, that we cannot be sure even of a rude approach to its total value, by attempting to add together a few of its nearest terms. We cannot tell, by such means, whether it may not prove divergent, so that negative terms of greater amount may render futile our poor attempts to find its approximate value. And the infinity is not of a single, but of

THE DOCTRINE OF UTILITY. 277

a double and triple kind. We have to trace out the results of the proposed action, not for a few hours or days only, but through a whole lifetime, or to distant generations, and throughout the life to come. We have to sum them up, by the theory, not with regard to ourselves alone, but to the whole family of mankind, and even to the countless numbers of generations still unborn. We must further trace them in connection with the immense variety of possible pains and pleasures, and their degrees of intensity. Each of the fifteen classes, which Bentham has enumerated, admits clearly of an almost countless diversity, not only in the strength of each conceivable form of pain and pleasure, but in the elements out of which they arise, and which must vary, more or less, with the moral antecedent which the problem requires us to determine.

The summation required is not only of an infinite series, with a threefold infinity of time, of persons, and of elements. It is also one of quantities wholly incommensurable. In geometry we may form a sum of numbers, or of lines, or of surfaces, or of cubical space. But we cannot form a sum of numbers with lines, or of lines with surfaces, or of surfaces with solid space of three dimensions. In each case a wide chasm of unlikeness or infinitude separates the proposed elements from each other. And in the moral problem, as proposed by utilitarian theories, the difficulty is just the same. It is owned, by one of the latest advocates and revisers of the system, that pleasures may differ in quality as well as quantity, and the admission is said to be quite consistent with the maintenance of the general system. The concession is candid and just. But the apology which attends it, for a master and teacher of logic, is most illogical. The essence and foundation of the theory is that the rightness or wrongness of actions must be determined by a summation of all the pains and pleasures which they generate, or to which they lead. But if these pleasures are owned to differ in quality as well as mere amount, the problem is either owned to be impracticable, or else completely changes its form.

Incommensurables are of two classes. The first is of things of the same kind, when the ratio cannot be strictly given by finite numbers, but we can approach to it as near as we please. In such cases, though scientific exactness is not reached, we can approach nearer and nearer to the truth, and a practical solution is possible. The other case is of things incommensurable in kind, as of lines and surfaces, or surfaces and solids. And here summation is strictly inconceivable. The gulf cannot be passed, and the only summation possible is by leaving the less important out of the calculation altogether. Now if the animal pleasures of sense, which swine may share, are owned to be distinct in quality from those of reason, thought, and intelligence, which are proper to men, the same principle clearly applies. And again, if the pleasures of lust, of selfish ambition, and of malevolent passion, the first and sixth and ninth in the utilitarian catalogue, are distinct in kind from those of justice, piety, and benevolence, here also a process of mere summation is impossible, and the attempt delusive. The pleasures and joys of a pure conscience are higher and holier than those of mere intellect, where no directly moral element appears; and these, in their turn, are nobler than mere animal pleasures. Hence it follows that the moral elements of the problem must first be proved to be in equilibrium, before we have any right to compare the merely intellectual gain or loss; and these, again, must be strictly equal, and disappear from the problem, before we

THE DOCTRINE OF UTILITY. 279

can lawfully introduce the pleasures of sense to turn the scale. Such a conclusion results demonstrably from the concession that is made, and it is fatal to the utilitarian theory. The addition table, before supreme, comes to occupy a very secondary place in this great problem of life; and the old distinctions of higher and lower aims, and the duty of choosing the highest, recognized by the best schools of heathen philosophy, as well as by the maxims of Christian faith, but set aside almost contemptuously by some modern writers, vindicate their reality and importance, and stand out in full relief once more.

But the problem proposed has still another fault, which is incurable. The calculation required is not only of infinites and incommensurables, but also of indeterminates. We are called upon to solve an equation of a degree which almost transcends the power of numbers to express, and of which not only the quantity itself to be determined, but nearly all the coefficients by which we must determine it, are unknown. Our calculation is to be of the supposed results of each act, as it will affect the experience of ten thousand thousand moral agents or sentient beings. And first we have to assume, in each instance, that they also calculate a like problem in each successive act, or else that they are guided by impulse, habit, or some lower motive, without calculation. In the former case the problem is clearly insoluble, and the mere attempt to resolve it is a childish folly. Each of a thousand unknown quantities is to be determined, by first assuming all the others to be already known. A is to decide whether a course of conduct is right by its beneficial effect on the experience of all from B to Z. And B, in turn, is to act on a similar calculation of the future feelings and experiences of A, C &c. to Z, as before. If we assume that others do not calculate, but act on mere impulse, the difficulty is only increased. We have not merely to determine, in this case, the future course and orbit of ten thousand moral planets, of all degrees of size, and with no common centre, under laws of high complexity; we have rather to decide, by calculation, the best course for one of the atoms of Epicurus, from the future effects on a wide chaos of like atoms, all assumed to move to and fro by mere caprice, or of which the secret laws for their movements are quite unknown.

But the calculation has still another fault. Instead of providing a sure and firm foundation on which the science of morals may rest, it builds plainly on the sand. It assumes that, while the voice of conscience may deceive, and common sense be merely a disguised name for despotic and arbitrary decisions of mere local or transient prejudice, this appeal to pleasures and pains, produced by certain acts, and the attempt to reckon up their amount, can provide us with a clear and fixed standard, free from the parallax and distortion of individual minds. But this is an entire illusion. It is a fact of experience, no doubt, in each case, and for each separate person, that such and such things have been liked and given pleasure, and that such others have been disliked, and caused an emotion of pain. But how many steps have to be taken, before this limited fact can be changed into a basis of scientific, or even of practical and approximate calculation! The pleasure or pain of the moment passes away, as soon as it has been felt, and can never return. How are we to infer future pleasures, through long years to come, from those of previous days? The circumstances will not be the same. The scene will have wholly changed. Every person af-

THE DOCTRINE OF UTILITY. 281

fected by the acts which are still in suspense will be in a different natural, intellectual, moral position, and the change is in progress hourly, even while the deliberation is made. Man, as we have been told long ago by the patriarch, "never continueth at one stay." The pleasures of infancy, of early childhood, of youth and manhood, are not the same. The choice of one is the dislike of another. The preference of to-day is often reversed to-morrow. Fulness breeds satiety. Vicious indulgence perverts the faculties themselves. And thus to build a theory of morals on a process of calculation, which assumes the nature and extent of future pleasures, in ten thousand persons, from our own experience, through a few days or months or years, of pleasures that are now gone for ever, is the same as to rear a stately and imposing structure on a foundation of ever-shifting heaps of sand.

To obtain anything like solid footing, we must turn our thoughts from the mere fact of some past pain or pleasure, of the liking or disliking, enjoyment or suffering, to the causes out of which it grows, and on which its recurrence must depend. The pleasure or pain results from a certain relation between the sentient being or the moral agent, and the conditions, natural or moral, in which he is placed. If there be such a thing as a right or wrong state both of body and mind, of natural or moral health, and of natural or moral sickness and disease, then the future pain and enjoyment must depend on both elements, the acts themselves, with their own proper tendencies when perceived by those who are in health, and those abnormal and unnatural results which flow from the presence of disease and moral evil alone. The first step of genuine science must be to distinguish these two elements, instead of blending them confusedly together. It must sever between consequences which properly belong to the action itself, and those which are due to the disease, ignorance, vice or folly, of those who come within the range of its influence. And when these last have been separated and put aside, as foreign factors which only obscure the real problem, those which remain must be classed under their three main heads, distinct in kind from each other, and also incapable of being fused into a process of arithmetic, because they are incommensurable. There are the pleasures of bodily health, in the exercise of all the outward senses and corporeal powers; the higher enjoyments of intellect and mental activity; and the highest of all, which include a moral and spiritual element, the delights of a conscience at peace within, and of a heart enlarged with love to mankind, and still tending upward to lose itself in the ocean of a love which is infinite and divine.

There is one further truth, familiar to every Christian mind, which places an insurmountable barrier in the way of any theory of morals, based upon consequences and results alone. We are thrown at once upon two alternatives. We must strive to determine, either the natural tendencies of the action, which may be thwarted and opposed, and partially reversed, by the wrong conduct, the error and ignorance of others, or else the results, which, in the light of former experience, are most likely to follow. If we are to deal only with tendencies, and not actual or probable results, we have then to decide what is the tendency which belongs to the act from its own nature, apart from those deflecting influences from other sources, which may turn it aside from its proper aim. And these tendencies must be of the nature of internal laws or moral forces. They must imply moral features of the act, and depend upon these, and can exist on no other condition. Kindness tends to beget kindness. But if the act of seeming kindness ceases to be really kind, the moral power is wanting. If it ceases to be felt as kind, the moral power is there, but ceases for the time to operate, through the ignorance, misconception, or ingratitude, of those from whom thanks are due. Instead of groping our way along the countless terms of an infinite series, we thus detect the secret law on which its development is mainly suspended, though temporary causes may obscure it for a moment, and introduce foreign terms that disguise its nature. To borrow the language of pure science, we deal no longer with the series itself, but with some generating function on which it depends, in which moral features are already present, and where spiritual laws shine out clearly, so as to rule over the consequences which flow from them even to distant generations.

But if we accept the other alternative, that our aim must be to determine the actual results, and not the tendency alone, then the doctrine must be set aside for an opposite reason. The improper results of actions, from the error and vice of others beside the moral agent himself, are various, confused, and infinite in our actual world. The attempt to trace them out in detail must defy the efforts of the most powerful and intelligent mind. But the process fails in another way. The world, in the faith of the Christian, and of every Theist, is not abandoned to chance or fate. It is under a scheme of moral government. The actions, then, of evil men are overruled by an Almighty Power. They are thus compelled, even against their own tendency, to minister to some great and worthy end of the Divine government.

Astronomy has supplied us, of late, with an expressive illustration and parable of this great moral truth. The comets, those wandering and unformed stars, like truants of our system, have been seen repeatedly, as they approach the sun, to suffer violent agitation. Jets of vaporous light have been thrown out from them, as if in weak rivalry or vain defiance of the mighty luminary which controls their motion. But these, however violent their agitation, are soon turned backward by a mightier force than caused their first eruption, and only increase the magnificence of the spectacle, while the whole obeys the central force that draws it nearer and nearer to the sun. So all evil actions, in their own proper tendency, seem to fight with the order and laws of the Divine government, and threaten to reverse them. They burst forth, in the form of violent and hateful passions and appetites, from that weak and errant centre of evil thoughts in which they are born, the human heart. But no sooner do they escape, and enter on the vast open range of Providence, than a mightier Power seizes upon them, in spite of their reluctance and opposition, and compels them to do its will. They no longer obey the impulse of the evil passion in which they had their birth, but are strangely overruled by far-reaching might and far-seeing wisdom, and minister to the action and abiding dominion of All-perfect Love.

Now this great fact, dimly taught even by natural reason, and more clearly revealed in the whole course of Scripture, that evil is overruled for good, and that nothing can escape from the counsel of One who is infinite in wisdom and goodness, destroys all contrast between good and evil actions, as soon as we adopt the false principle which would decide on their character by actual results and consequences alone. In this sense the words of the poet are true:

> All discord, harmony not understood,
> All partial evil, universal good.

All good actions, in their own nature, tend to results which are good. And this tendency, inherent in them from their moral nature, is ripened and perfected by the concurrent action of a superintending providence. All evil actions, in their own nature, tend to evil. But this evil, the result of their moral nature in the hour of their birth, is so controlled by that higher law of Divine wisdom, under control of which they come as soon as they are born, that they also lead to actual issues worthy of Him whose decrees they fulfil,

> Whose work is without labour, whose designs
> No flaw deforms, no difficulty thwarts.

How, then, can it be possible to discriminate the moral character of actions by their outward results, when all alike, though in a manner deeply mysterious, are known to minister to a work which is perfect; and when God was pleased to accomplish, by means of an act which on man's part was one of the worst and foulest crimes, His own good and excellent purpose, the redemption and spiritual recovery of a sinful and evil world?

When, however, instead of making the tendencies of actions, which is a silent paralogism, or their actual results, which is a fatal error, the ground of their moral nature for good or evil, we use the lessons of experience as a secondary guide for settling the wisest forms and true limits of human law, then the principle of utility, so applied and limited, must hold an important place in all applied Ethics, and needs to be borne in view by all those who have to frame laws for the guidance of human society. Or when we rise still higher, to contemplate the Providence of God and the great system of His moral government on a wider scale, and seek to define good and evil by the harmony and consent of all those contrasted elements on which their nature depends, the principle holds a place of still higher importance. We cannot sever actions from their consequences. If the tree be good, its fruit will be good, and if the tree be corrupt and evil, so must be the fruit also. A perfection, wholly apart from consequences and effects of every kind, is inconceivable.

Actions that are really good must have a threefold character. They must be good in the fountain from which they spring, the channel through which they flow, and the ocean to which they tend. They must be conformed, in their origin, to eternal laws of righteousness, on harmony with which their real character depends. They must be in harmony with a pure and upright conscience, which recognizes the law of right, and strives to obey it. And they must tend naturally, in their later progress, to reproduce themselves. They are moral plants, endued with a prolific virtue, and of each of them it may be said with perfect truth, that "its seed is in itself upon the earth."

Heroic virtue wakens emulation in noble minds. Holy examples awaken the desire and thirst for holiness. Righteous government kindles instincts of justice, and diffuses them through the whole nation. Kindness breeds kindness, and love is the fruitful parent of love. And though actions could have no moral tendencies, if ethical distinctions were set aside in the vain search after some simpler theory, and right or wrong doing were to be defined by visible success, and measurable results alone, still it is often through careful observation of results that our perception of their moral nature becomes more full and clear. Like sunlight reflected from the countless dewdrops of morning, the laws of righteousness, while they can gain no new force, shine with more conspicuous beauty, when they are seen in their combination with all the innumerable events of Providence and human history in this lower world. Actions do not assume a moral character, and become right or wrong, purely because of the results that follow after them in the near or distant future. But the great fact of an all-wise Providence secures that, soon or late, their moral tendency, from their own nature, shall prevail over every adverse and evil influence;—" so that men shall say, Verily there is a reward for the righteous: doubtless there is a God that judgeth in the earth."

THE ANALOGY
OF
MATHEMATICAL AND MORAL CERTAINTY,

A Commemoration Prize Essay

READ IN TRINITY COLLEGE CHAPEL,
DECEMBER, 1833.

B. L.

AN ESSAY

ON

MATHEMATICAL AND MORAL CERTAINTY.

AMONG the votaries of Mathematical Science there have never been wanting voices to celebrate largely its numerous excellencies. The simplicity of its principles, the variety of its results, the certainty of its demonstrations, the beauty and harmony of its truths, and its wide command over the secrets of external nature, have all in their turn been the themes of frequent eulogy. Regarding it as the parent region of discovery, not only its native children, but many of the sons of tributary sciences, have been ready to pay their homage to its superior claims. By all who have journeyed in quest of physical truth its praises have often been sung, as mistress of those winding seas they have had to traverse; and in the unknown lands of science which have risen on their view, its flag has been the first to wave in triumph upon every shore. Others have claimed for it a higher merit; they have enlarged, and justly, on its influence upon the mind itself —on the habits it tends to induce of connected reasoning and patient thought—the strict rein it places on the loose wanderings of the fancy, and that stern discipline of the mental powers, which nerves them equally, whether with strength for the combat, or with swiftness for the race.

But here the topics of praise seem to be exhausted, and the strain, so full hitherto, dies away suddenly on the strings. To those, who, deaf to its lower claims, inquire into its moral virtues; who adhere to the golden rule of the reviver of philosophy, and measure the value of science, not by the false standard of that power over the elements, which spirits of darkness may wield as well as spirits of light, but by its use in the conduct of life; who ask, what higher truth it even indirectly teaches, what light it sheds on the mysteries of our being, what lessons of wisdom it has power to furnish, what sympathies to cherish, what kindly affections to revive, what lofty purposes of excellence to animate and sustain, the reply, if any reply be given, is faint and low. Its more just and candid admirers will excuse, perhaps, by the greatness of its other merits, its defect in these, as alien from its nature; while some of a bolder school, the idolaters of science, who dream that wisdom itself may in time be reduced to a formula, and virtue to a refined and subtle problem of chance, will treat the inquiry with contempt and silent scorn. The truth, however, of the charge would probably be at once admitted by both; and while all other fields of natural knowledge, to those who have learned to pursue them aright, yield in abundance the fruits of lofty poetry and of heavenly contemplation, it seems to be deemed utterly hopeless that any such plants of Paradise should be reared in this primitive region of granite rock and eternal snow.

The source of this wide-spread impression seems to lie in the opinion, which has taken deep root, of an entire contrast between the nature of mathematical and moral truth. They have been generally viewed as not more separate in their objects than in the kind of evidence on which they severally rest. The claim of demonstrable certainty has been granted to the former alone, while the latter has been placed on the looser ground of induction and cumulative proof. The great and leading distinction, which obtains, by a close analogy, in things moral as well as natural, between facts and truths, the floating phenomena apprehended by the senses, and the fixed laws or ideas contemplated by the reason, has been transferred, by a strange confusion of terms, if not of thought, to the subjects themselves, till the very phrase, 'moral certainty,' has come to denote what is in some measure uncertain, and not capable from its nature of the strictness of demonstration. Between things regarded as so different in kind it is no wonder that little analogy should be felt to obtain. On this view the only lesson that could be drawn from the truths of science was one of contrast, to seek for such evidence only in moral subjects, as their nature admits; to balance where we cannot prove, to creep where we may not soar, and steer our winding course along the coasts of experience, where there is no fixed loadstar seen, to guide us freely across the trackless ocean.

There is something in this opinion, when entertained, which must, to a reflecting mind, suggest itself at once as strange and unnatural. Is it possible, then, that nature, which in every other province has a temple, an oracle, and a voice of wisdom, in this alone, the highest of all, is mute and silent? Her smaller streams and fountains, amidst their perpetual dance and play, may be seen lit up continually with the beams of some loftier truth; and this ocean, wide as space, and lasting as time, which encircles and sustains every realm of the universe with its vast and eternal laws—has this alone no clear heavens bending over it, whose image it may reflect in its deep and tranquil bosom? Is Truth, as distinguished from Presumption, her shadow, to be exiled from the spirit of man, and the nobler aims and purposes of his being, to seek an uneasy shelter, entrenched within the lines of the geometer, or couched in mystic retreat beneath the letters of the analyst, there to busy herself with many things, and all needless, ministering only to bodily convenience, and outward luxury and ease.

Let us briefly inquire, then, whether we may not with safety take a nobler view; whether, in examining into the nature of moral certainty, we may not find, in the certainty of geometry, an attendant to guide us, instead of a rival to discourage; whether we may not rest their claims on a common footing; observing only that along with the mental discipline they both require, in the former a moral discipline is further needed, answerable to its greater dignity; and thus rise to the conviction that, in the range of moral as of natural truth, there is no need to be tossed on the uncertain waves of opinion, but that in each alike there are principles to be found, fixed and sure in themselves, and whose light streams far into the caverns of nature, and the still more unfathomable depths of the human heart. And although the remoteness of this view from those of the schools of modern philosophy, with a few bright exceptions, may give it the air of paradox, it is not on that account the less likely to be true. For what is paradox, indeed, but another name for that deepest truth, which veils itself from a passing and careless gaze in the inner shrine of this world's solemn temple?

At the very entrance, however, of our comparison, a host of difficulties encompass us, and threaten to obstruct our path, like the clamorous shapes which beset the knight of old, when forcing his toilsome way to the palace of Truth. Their form and voice may be various, but their challenge is the same; and as they point to the countless train of varying thoughts, adverse schools, contending creeds, and clashing opinions, they ask with triumph where are the signs of this fancied certainty, the doctrines of this science, the waymarks of its progress, and the treasury of its results; and setting before us the trophies of physical research, demand an equally undisputed array of moral and spiritual discoveries.

It cannot be denied that, at first sight, this difficulty seems great and almost insurmountable. The analogy appears buried under a total contrast. And so it must appear, as long as the eye rests only on the passing surface, unmindful of the still deeps of thought that are slumbering below. To such an observer, and such are most, the history of the world presents only a waste of opinions, where wave and tide and current blend their effects confusedly, hopeless to the feet, and cheerless to the view. And even when the voice of Truth is heard to welcome him amidst its waves, his spirit must be calm and his faith be strong, who can walk forth in thought, without some secret misgiving, over this wide and wildering sea. Yet, however formidable the difficulty may appear at first, further reflection will greatly weaken its force, and supply us with a double answer, resting in part on the analogy in question, and partly on the difference between the circumstances of scientific and moral research.

Let us suppose the case of a simple observer of the heavens, a Chaldean shepherd on the plains of Dura, who had watched night by night, the moon, the planets, and the starry courses; and that in the midst of his laborious observation, still finding fresh aspects in the ever-moving sphere, he were told that all these varied appearances were linked by one simple law, from which they might be deduced with infallible certainty. We may imagine how he would start at such a statement, how he would appeal to their complex and still varying motions; and gazing on the mystic dance of those wandering fires, "excentric, intervolved"—be ready to seat some wayward Genius in each separate sphere, and hard would be it to persuade him that they were "regular then most, when most irregular they seem." But if by comparing with other observers, he were to learn that with each the appearances were different—that stars which to him rose and set continually, by another were never seen, and with a third were circling as constantly, unwet by the ocean wave, the discordance of the appearances with every observer would appear to give him a tenfold warrant for rejecting the statement altogether, as rash and visionary.

Thus it fares with higher truth. Habit and prejudice, the pride of intellect, and a selfish will, furnish to every man a continual parallax of thought; and till these are taken away, or account is had of their presence, we need not wonder that the results appear discordant, and no light of certainty pierces through the gloom. Remove these, and what before appeared the most complex may be found perhaps the most simple. In this view it is worthy of notice that the first triumph of applied analysis has been over those bodies, whose very name marked them out to us as truant wanderers in the sky. And if Light, the playmate of the changing clouds, be the next to bow to that potent spell, what presumption can hereafter be raised, from the complex and tremulous emotions of the heart, the web of opinions, or the tangled maze of the world's history, against the simplicity of those eternal truths on which the whole scheme reposes, or the sureness of the building reared on those foundations?

There is, however, a further key to this apparent contrast. The geometer never dreams of discovering afresh the whole train of science: he receives it advanced by the united labors of all who have gone before, and seeks only to supply what may be wanting, or explain what may be obscure, and so to pass on to further discoveries. He would smile at any one, who would advise him to ensure freedom from prejudice, by flinging aside the treasures that others have left, and raise the structure afresh by his own unaided powers. If this had been the course pursued by even the most favoured sons of Science, the walls on which she now hangs their trophies would still have been bare. Yet something like this has been too often advised, and too often practised, in themes of far more difficulty and moment, and it is no wonder that the progress, in such cases, should appear small and uncertain.

Above all, the difference lies mainly in the number of willing learners, rather than in the subjects themselves. There is something in the pursuit of pure science not a little flattering to the mind, and tending to leave on it a simple impression of its own power. With moral truth it is the reverse; it humbles that it may exalt, and reproves first, in order to restore. There is a sacrifice always involved in its reception, a surrender of selfish desires, a deadness to passing pleasures, which bars the avenues of its approach, though it only confirms it the more when once received. This, while detracting nothing from its intrinsic certainty, which it rather doubles by the joint testimony of the judgment and the will, clouds greatly the marks of its evidence from the superficial eye.

Thus many turn altogether from its portals; and even with those who enter in, it is frequently only after laying afresh, in their own experience, the foundation of a wisdom taught and proved long before.

It is true that this view of Moral Science, as in its own nature fixed and sure, like the laws of number or of space, however familiar with our elder writers, and ably sustained by some kindred spirits in our own day, accords ill with the stream of modern opinion, or the self-flatteries of an age, which, whatever may be its other intellectual claims, shews small signs of the unity which springs from conscious truth, or the meekness of a pure and living wisdom. And while its merely physical students or ephemeral writers keep within their proper bounds, they may well be borne with, and even justly admired. But if they would intrude, with insect doubts, into the still chambers of meditation and faith, to question the existence and reality of that higher knowledge, which they may never have sought for, and therefore have never found, their pretensions must be examined, and the intrusion repelled. And it may easily be seen that the causes which obscure the evidence of moral truth, though always acting, act most strongly in times of intellectual excitement and social luxury.

We may observe then, first, that a state of accumulated facts, multiplied opinions, enlarged materials of thought, diffused intelligence, and various reading, is of itself the most unfavourable to fixedness of views, and manliness of character, unless it be attended with an answering depth of inward research and silent meditation. Truth, and the facts which it knits together and explains, like law and its subjects in the social state, or flesh and spirit in man's mysterious being, are of themselves antagonist principles; noblest indeed in their union, but still that very union is not effected without a struggle, and the harmony which results partakes always the nature of a triumph. The appearances which flit around us in every field of nature are of themselves a vagrant and lawless band, till the idea, in full panoply, springs to birth in some master-mind, which may reduce into unwilling obedience their roving and scattered tribes. And hence, in every case, the greater the amount of facts amassed, before a sound theory has arisen to explain them, the greater the obstacles which such a theory will have to encounter. The only reason why this is not perceived more commonly, is, that facts and theory usually grow up side by side, and advance together towards perfection. But if, as in the formation of opinions on moral subjects, where the exercise of a private judgment, though sincere in its search, will too often set aside authority both human and divine, principles have to be formed and matured amidst the pressure and excitement of intellect, in its most busy and active forms, then the difficulty exists in all its power. To persuade this moving and infinite world of thought to own the jurisdiction, and bow to the authority of a higher law, fixed and stable as those of number and space, is a task which grows the more arduous, the more populous that empire of thought becomes. Let us suppose no longer a shepherd on the plains, but a modern observer of the heavens, who had collected the observations of former astronomers, and added his own, with every refinement of minutest accuracy, before the first dawn of a true theory. It is plain that with such a person the difficulties, which a later Kepler or Newton would have had to encounter, would be far greater than those which met them at the time when they actually appeared. Few and loose comparatively as were the materials then, and rapid as the advance of science had been in the preceding century, the latter had to task its power to the utmost, to achieve that first great victory, in the natural world, of law over experience, and reason over the illusions of sense. How much harder would it have afterwards become, when a thousand disturbances then unnoticed and unmeasured, rebel genii of air and earth and heaven, would have been added to the array, and have seemed to mock every effort to bring them under the yoke of theory! And perhaps at last the results themselves might perish, from the want of any principle to knit them into union, and give them a voice prophetic of future changes.

This, which is true of the laws of matter, is true also of the laws of mind. Each alike has its region of pure and applied science, the one, sure in itself, and the other scarce less so, while by the aid of some simple fact assumed, it explains, in strict consequence, the varied forms of nature, and the forms, as various, of providence and of human life. And it can be no presumption against the equal evidence attainable in the latter case, that it may seem to be seldom reached, even in an age of great mental development. The man of various reading, if a man of various reading only,—the able and practised statesman, if versed only in the subtleties of state affairs, the traveller in many climes, if his own heart and conscience have been left unvisited, the citizen of the world in every form, if a citizen of this world alone, is the least likely to apprehend the evidence, realize the depth, or trace the far-spreading consequences, of those laws of judgment and righteousness, which are the foundation even of the throne of God. Every rule in their mind is clogged with exceptions, which leave it worthless, every maxim with questions and doubts which destroy its power; and each guiding star, that might beam down benignant influence, meets there with some gross medium of earth, to distort at least, and perhaps entirely to quench its ray. And thus an intellectual age, from the multitude and weight of its materials of outward knowledge, is only in danger of sinking the deeper into the quicksands of scepticism and doubt on every truth of the conscience and reason, beyond the boundary-walls of time. On the first whisper that here too the certainty of demonstration may be gained—the troops of its jarring opinions, which, under their several factions, have striven long and in vain for mastery, muster at once on their frontiers to resent this intrusion of glimmering light into regions sacred to strife and dusky gloom. Freed then, in some measure, from the fear of this objection, which flings its shadow across our path, we may pursue our inquiry with less doubt of success, and trace rapidly the leading steps of demonstration in either case, illustrating by comparison the difficulties that arise as we proceed.

Every stream must have its fountain, and every course of demonstration some higher principle, out of which it springs. These, which are the ἀρχαὶ of Aristotelian philosophy, form the root of science—the postulates with which all reasoning begins, and without which, indeed, a single step cannot be taken. They are of two kinds, primitive, or dependent; according as they spring from the intuitive laws of thought, or are the result of a laborious induction. The former are the source of pure, the latter of mixed science, and the certainty to which they tend may in like manner be termed pure or mixed, as the principles on which it rests are absolute or dependent. That amusing travesty of logical forms, which

Dr Brown has copied from Pope to raise our mirth at the Stagyrite's expense, contains in it, as with ridicule is often the case, the germs of a deeper wisdom than either seems to have suspected. Truths, as language itself may teach, have their families, their order and succession, their parentage and their offspring: the child cannot usurp the place of the parent, nor the parent of the child. Unlike the shadowy forms of mere opinion, they are endued with an intense vitality, and their incorruptible life multiplies itself a thousand fold. But as there can be no family without its patriarch, and no race without its founder, whose breath of life is the immediate gift of heaven, so have all the applied sciences their respective parent truths, and the pure science from which they spring its intuitive principles undemonstrated and underived. And thus is man taught everywhere the same lesson of looking upward, and the only tenure by which he can till the fields of science, is the silent acknowledgment that Mystery is empress of the soil.

Overlooking, or strangely forgetting this simple and evident maxim, some writers on mental philosophy have perplexed themselves at the very threshold of their inquiries, one class attempting to set aside as incapable of proof, and others as idly seeking to prove, principles, which, whether they were willing or not, they were compelled silently to recognize in every page. And perhaps even the labour might have been spared, however ingenious, which others have spent in proving that no proof is needed, and that the lecturer may safely assume the mental identity and bodily presence of those whom he is addressing. Discussions of this kind of themselves may seem curious and harmless; but the habit they foster of questioning what is unquestionable, and perplexing the consciousness with the attempt to contradict itself, tends greatly to induce a palsy of the soul; in which it grasps no truth firmly, has nerve for no manly purpose, but trembles with a general uneasiness of doubt; and the sense of duty is lulled asleep, while it is gravely resolving the question of its own existence, and of the living, moving world around it.

The principle, then, which lies at the basis of mathematical science, is this consciousness, which our senses furnish, of the existence of an outward and material world. On this simple postulate, it rears its various discoveries. Thought and reflection may suggest on this subject numerous inquiries; many curious, and some perplexing questions may be started, on the process by which the mind learns to clothe one object with the perceptions gleaned from differing senses. But these inquiries can never shake, nor even modify the conviction itself. It is just as strong in the sceptic philosopher, who strives to melt away his own consciousness in a *deliquium* of ideas, as in the merest peasant, and thus gives us a sure footing on which to rest in every further course of reasoning.

And yet this conviction, however simple and sure, may serve to remove the objections which have sometimes presented themselves in the parallel case of moral consciousness. For it is one which we hold in spite of many seeming difficulties. Even in the fields of nature, and, still more, in the chambers of art, we meet with visual images that mock us with an unreal form; and sleep too has its own world, filled with sights of beauty that own no outward archetype, and vocal with sounds of love or friendship that were never uttered.

All moral science, in like manner, springs from a fountain-truth, as simple, and even more deep in its intuitive certainty—the inward consciousness of Will, and that fixed sense of responsible being, which at once proclaims the soul a denizen of the world of spirits. It is, like the former, a primary law of thought, which needs no proof and requires no comment, but has its quiet and undisturbed abode in the inmost chambers of the heart. It embodies itself in the pronouns of every language, breathes in every desire of the soul, lives in the memories of the past and hopes of the future; and its voice is heard with equal clearness, in the infant's lisping confession, or whispered prayer on its mother's knee, as in the profoundest maxims of grave and hoary wisdom.

It is true that here also there are difficulties, which only patient reflection can overcome, and which shade the truth though it will not suffer itself to be buried. Even when in its most wakeful hours and purest contemplations, the mind may sometimes gaze, Narcissus-like, on the images of its own creative fancy, till it is ready to confound the shadowy with the true. Still worse, the soul too has its sleep, and like the hero of Italian song, in the enchanted isle of pleasure, with witcheries of delusion on every side, lies too often a dreaming and inglorious captive, deaf to the sounds of wisdom, and dead to the visions of immortality, that solicit an entrance to its ear and its eye. In such a case, it would be as vain to look for any certain discernment of moral truth, or any clear insight into its lessons, as to expect the theorems of geometry to evolve themselves spontaneously in the slumbers of the peasant. Rather may we well imagine that a dim and uncertain haze will creep gradually over the whole subject, till first principles themselves are obscured, and a darkened understanding 'makes one blot of all the air,' with its subtle spells. As one who has long been drugged with opiates, and seldom opened his eyes to the light of day, might almost learn to doubt the evidence of his waking senses; so by such dreamers as these the very existence of moral distinctions, of right and wrong, of good and evil, comes at last to be mooted. Yet even here, the unconquerable majesty of Truth asserts its power. In the midnight haunts of dissipation, where riot holds its loudest orgies of uneasy laughter and hollow mirth, the roar and revelry will sometimes be hushed and still, and a voice be heard of clear and solemn music, that bespeaks something nobler than 'any mortal mixture of earth's mould'—the plaining of a spirit parted from its guide, and lost in the entangling wildernesses of the world.

There is no reason, then, in the nature of its first principles why the evidence of morals should be placed on a lower footing, or regarded as less purely demonstrative, than that of geometry. Each rests on its own firm and simple basis, in the instincts and intuition of sense and of conscience; and their difficulties are similar, though not impairing the certainty in either case, for sense has its illusions, and conscience may often be obscured with clouds. With advancing knowledge, however, these difficulties, which were perplexing or startling at an earlier stage, yield fresh confirmation to the realities, which they seemed before to involve in doubt. The distorted vision, and spectral image, are thus traced to the same laws of light, which serve to link us with outward matter; and false and varying standards of human morality confirm us the more in that which is pure and divine, when once we have learned the mysterious power of a fallen will, to bend with its mist the light of conscience from that unbroken course which it holds in mild regions of serene and heavenly contemplation.

When, indeed, we pass from the nature of Moral Science in itself to its practical reception and growth, we can scarcely allow too largely for this last influence. The doctrine that belief is unconnected with the will may pass current with those whose horizon is bounded by the rules of arithmetic; in moral subjects it will not hold. And hence it arises that the outward marks of certainty or progression are, to the casual observer, so widely different from those of natural science. The strong sense of this actual disparity, has perhaps led a well-known living author, in his aphorisms, in some degree to overstate the real distinction, when he says: "The postulates of Geometry are such as no man *can* deny; the postulates of Moral Science such as no good man *will*." And this is the view which does seem most directly to suggest itself, upon observation of the wayward forms of human thought; but perhaps a closer reflection would lead us rather to state the difference thus—that the postulates of Geometry, no one denies or wishes to deny—those of Moral Science, they who wish it most earnestly cannot always. This view will explain the practical distinction as fully as the former, and the parallel is seen to hold with undiminished clearness.

In pursuing the course of demonstration, the next step is to separate the abstract relations which present themselves from the varying circumstances in which they are involved. To effect this, simple and precise definition is required. And here the need of patient attention begins to be felt, and we tread close on the verge of mystery. We have now to contemplate the relations of space, apart from the forms of matter, by which space itself becomes known to us, and the nature of number, distinct from the qualities of those units by which our ideas of number have been gained. The figures defined by Geometry cannot be viewed as merely states of the mind, without losing their objective character; and yet we cannot ascribe to them objective existence, without encumbering the idea with the addition of specific size. They seem rather to be viewed as the expression of general relations which the mind apprehends in external objects—relations fixed and unchanging, though the objects vary, and which the thoughts invest with a subjective or ideal unity. But though the process is not the most easy to explain, it is one which every mind, whose attention has been closely fixed, pursues naturally, and without difficulty. And when once the relation, whatever it be, has been strictly defined, its properties are felt to depend no longer on the object or diagram that may aid our conception, but to be conclusions of pure reason, resting solely on what the terms of the definition include.

The path which opens into the fields of ethical knowledge is of the same kind. The actions of human life, the varied emotions of the heart, and all the shifting scenes of thought and passion, are set before us; and the first impression, as in the natural world, is that of a multitude that distracts, and an immensity that confounds. The unthinking turn away discouraged, and live, as the rustic amidst the scenes of nature, but with a more dangerous ignorance—convenience or custom their only guide, their thoughts cooped always within the little horizon of their own narrow aims; and if ever they raise their eyes, as at times they must, to the blue heavens, yet soon turning them down again to earth, contented strangers to the meaning of all those high mysteries, whose silent majesty is encircling them evermore. But those who with persevering attention continue to gaze, will ere long perceive ideas of the good, the fair, the lovely, and the pure, offering themselves unbidden to the view, and amidst the innumerable play of life's changing scenes, rising, like the nymphs from the countless rippling of the ocean waves, in forms of immortal and unfading beauty. And though writers of an older school, whose thoughts dwelt much and long in this sphere of contemplation, have naturally used words answering to the vividness of their own impressions, and spoken of good and evil, of vice or virtue, in terms that might imply a separate actual existence, it would be wrong on that account to charge them with entangling themselves in the empty fictions of a creative fancy. Language, in spite of ourselves, will clothe the most abstract thought with some floating drapery; and the same objection which would discard the certainty of moral relations, or dim the impression of their simple power, because always enveloped in varying circumstances, must fling aside the conclusions of the geometer, till he can set an abstract triangle or circle before our eyes.

It is in this stage of comparison, that the inquiry into the source and character of moral distinctions, which has employed so many writers, and been the theme of so many discussions, ancient and modern, would naturally present itself to our notice. For the practical guidance of life, the inquiry is happily superfluous: this temple, sacred to the charities and the graces, will not be less sure for the feet, or less goodly to the eye, because its foundations may be concealed in the bosom of the earth. Conscience has its voice for every ear that is open to receive it, and misleads none, who have not first been more or less willing to be deceived. But still the subject is in itself one of the greatest interest, and especially after furnishing the theme for such various disputes and not a few specious falsehoods, a just view of its nature may be deemed almost the first essential for all who would pass beyond the sphere of individual action and accept the more perilous office of the championship of truth, and vindication of the ways of Providence. Of secondary aids in pursuing this research, there is perhaps none so sure and simple as the light which arises from keeping the present analogy steadily in view. To attempt this generally, would lead us too far; it may be enough to illustrate briefly in a single instance; and the philosophical lectures of a recent author (Dr T. Brown) as among the latest who have treated the subject, may serve as a ground for the observation.

Agreeing then, with that luminous writer in most of his statements, there is one principle in which he appears to be still defective, and the error, though less obvious in its consequences, to be as truly dangerous as those which he has opposed. Laying his foundation more deeply than some of his immediate precursors, he had yet to dig a course deeper to find the rock on which the building might rest securely. He has swept aside with a masterly hand the cobwebs of the selfish theories—those fair-seeming Florimels of ice and snow, which but for the charmed girdle that subtle thought had placed around them, the first warm feeling of the heart would have melted away. He has further remarked that selfishness loses nothing of its nature by reaching from time into eternity, and seems only more heartless, when it stands unmoved in the presence of Infinite Goodness, and even in the hope of immortality thinks only of the gifts, and forgets the Giver. He has even gone further, and traced clearly the evil results of ascribing the source and ground of obligation to an outward expression of the Divine Will, instead of an inward feeling of the soul ;· however true it be that those feelings when unperverted, must secretly recognize that Will's supreme authority. Otherwise, indeed, it could possess no moral goodness, and the sole attribute of the Divinity would be Almighty Power. But when he proceeds, as he does continually, to speak of these principles and feelings, as purely of positive Divine appointment, which the same hand that implants them may change or reverse, he fails to observe that he is secretly retreading the steps of his own argument, and the giant falsehood he had just exposed, which strips even the throne of heaven of its glory, rears itself from its fall with recovered force.

The sense of good and evil, it is true, are prior to the expression of the Divine Will, which does not create, but only awakens them. This conviction, always felt, though sometimes a mistaken zeal may have sought to obscure it, alone leaves any meaning to that earnest enquiry, 'shall not the Judge of all the earth do right?'—That which defines cannot, from its very nature, become a property; and thus we cannot justly speak of that Will as good, so long as we regard it as constituting of itself the final standard of right and wrong. But it is plain that we are brought equally to such a conclusion, whether we view it as directly propounding itself as such a standard, or implanting feelings arbitrarily to which it afterwards refers. The path leads us by a longer circuit, but the issue is the same: indeed, when searched narrowly, the difference is only this, that a secret sense of deception, painful even to entertain for a moment, aggravates the evil of the former view. The contrast is the same as would be felt between the sway of an arbitrary sovereign, who should claim obedience to his commands on the sole ground of power; and one who should further claim the praise of his subjects for governing with the sanction of a senate of his own appointing, and tamely subservient to his will. Yet surely this would be a faithful emblem of the divine government, could we conceive the moral emotions as of arbitrary appointment, and liable to be reversed at will so that what we now condemn as evil, we might in such a case have been taught to praise as good. From such a view the mind recoils even more than from the former, and takes refuge in the parallel expostulation, 'Shall the throne of iniquity have fellowship with Thee, which *frameth* mischief by a law?'

Now if we accept the illustration afforded by geometry, the mist clears from around us, and these spectral difficulties haunt us no longer. The relations which there present themselves, as soon as perceived, suggest the feeling of unchangeableness. Though both the forms of matter from which they have been abstracted, and the mind which perceives them, readily acknowledge a creative will, with these the thought cannot be associated, without the sense of an entire disparity. The only aspect in which we can regard them, is as the objects of an all-including intuition, but not as the objects of arbitrary power. And in meditating closely on Moral Truth, we shall find the same conviction equally strong. It is as easy to conceive the laws of space inverted as the laws of conscience, and to grant that the properties of the circle might change, as that treachery should become a virtue, and compassion a crime. So rooted is this conviction of good and evil as anterior to choice, and law, as in thought at least earlier than will, that reason and revelation unite in tracing it even up to the throne of God, and

He who sits thereon himself accounts it his glory, that He delights in judgment, justice, and loving-kindness. And thus, in the words of Hooker, the "perfection which God is, giveth perfection to that He doeth," and that Eternal Law, whose voice is the harmony of the world, and to whom all creatures in heaven and earth do homage, finds in His bosom her seat and surest resting-place.

If any, however, shrink from this view, as seeming to affix bounds to Infinite Power, they will do well to reflect that, by rejecting it, the still nobler attribute of wisdom is entirely excluded. Wisdom lies in the choice of the highest ends, and the best means to ensure their accomplishment; and the idea of means and of an end alike involves a fixed relation, so that without the former the latter could not equally be ensured. Exclude this, and a series of effectless and unconnected acts comes in the room of moral government, and a cold and iron fate usurps the place in the thoughts of a widening and still progressive Providence.

There is another subject of much practical importance, which the present comparison may assist in explaining, and which will serve to account for much of the uncertainty often associated with moral science. In the barrenness of imperfect definitions, which describe loosely, without defining, we may learn the effect of a low and variable standard of right and wrong in impairing the vitality of truth, and blocking up the avenues of wisdom. It is not enough, for instance, that a triangle were defined as a figure bounded by three sides, unless we imply that the sides are strictly right lines. A single consequence could not be else deduced, and the figure might pass by insensible change into a tortuous curve. And very similar is the result, when we would bend in our reasonings the rule of integrity, however lightly, to meet some purpose of passing convenience. No further lesson can then be learned: the key is twisted from its natural shape, and will no longer open the mysteries of Providence, or move freely in the wards of the heart.

The last step which requires notice in all demonstrations is that course of deduction, which, starting from previous definitions, passes into all the varied theorems of science. To define is to lay the foundation, but this rears the superstructure. There are two modes of arriving at truth, 'induction,' and 'deduction,' or syllogism. In the former we rise from instances to general maxims, in the latter we re-descend, whether from the maxims of induction, or the intuitive truths of reason, to remoter results. And the figure has further propriety, for in rising, the course, though safe, is laborious and slow, while the descent is easy, but a single false step may prove dangerous to the whole.

It is amusing to observe, how in the schools of modern philosophy, as in other schools where the master has lost his authority, reprisals have been taken for former bondage, and it seems felt as the greatest triumph of their young liberty, to pull Aristotle or Plato by the beard. The treatment which the syllogism has received at their hands is a striking instance. It appears to haunt them with the memory of past subjection. Every folly of the schools, not to say every crime of the middle ages, has been laid at its door; and the vehemence with which they have exorcised it might lead us to suspect that this was the triple knot of darkness, which so long had shrouded learning in its gloom.

On a calm examination, we find little ground for this indignant contempt with which it has been greeted. The syllogism, as a recent encyclopedist has justly stated, was never designed by Aristotle for a new and peculiar degree of argument, and still less, as the rather flippant remark of Locke would imply, as a means for making men rational, but simply as an expression of the course which the mind silently pursues in all its reasonings. It is a method for spelling out distinctly the steps of an argument. A reader, who should spell every word at length, it is true, might make us first laugh, and soon grow weary. But though in correct and tasteful reading many letters are always silent, and even many words passed over lightly, still it can scarcely be doubted that he who would read without danger of mistake must at some time or other have learned to spell. And so it is in the case before us. A train of reasoning, as expressed, may usually be only a series of minor propositions issuing in the conclusion; and in the readiness of the mind to invent, and its strength to retain these, resides mainly its reasoning power; but the major must always be silently implied, or fallacy creeps in, and the argument fails. And hence, though in point of expression a syllogism might be called a redundant enthymeme, in point of thought it must be found that an enthymeme, as its name implies, is only a secret and mental syllogism.

Against the course of syllogistic reasoning, the writer before alluded to, in common with several others of equal note, brings forward a more important charge. They accuse it of entire and necessary barrenness. And as this perhaps is the most common source of the opinion that the certainty of science is not attainable in morals, it will require to be briefly examined. The charge is, simply, that a syllogism begs the point in debate, and assumes the very thing it is wanted to prove; and that, since the major involves the conclusion, the latter must be felt to be true, before the former can be justly admitted.

It is easy to see that this triumphant charge, which is brought with the air of a valuable discovery, does not lie against the forms of logic, but against the very nature of deductive reasoning. To deduce is not to create or to invent, but merely to unfold. The fallacy lies in supposing that when a proposition is felt and granted to be true, all its various consequences must also be felt and known. In a separate syllogism this may seem to be the case, and here is the secret of that fancied leanness to which Reid and Brown allude; but to denounce, on this ground, the whole process of deduction as unnecessary and superfluous, is as reasonable as to persuade a man that his legs are useless, because he appears just where he was after a single stride.

The present analogy, however, more palpably sets aside this objection, and places the importance of deduction in its fullest light. The demonstrations of geometry are simply a course of syllogistic reasoning; and though to throw the whole into mood and figure would be absurd, yet, on suspicion of a lurking fallacy, any part might be presented in a syllogistic form. Do we then speak of the process as idle, or the results as useless or uncertain, because they must all have been included, as truly they must, in the principles from which they have been derived? Or rather, is it not the greatest boast of this science, that from definitions so simple, and axioms so few, it brings to light conclusions innumerable, that spread into all the fields of nature, and supply all the arts of society with ceaseless inventions? For, to use again the figure which a misplaced ridicule supplies, it is not so much the elder sons of truth's immortal family, as its younger descendants, that stoop to intermarry with the dying phenomena around us, and yield the offspring of arts that minister to the comforts, or lessons that aid in the guidance, of daily life.

Now there is nothing in the nature of moral subjects to preclude our offering the same answer. The spirit of man has its own self-evidence, conscience its own definitions, and those not cold and silent, as in geometry, but vocal with strong emotion, and endued with a living power; its truths are in their nature as prolific; the opening for deduction is the same; and why should it not lead to certainties as sure, and conclusions as various? Even were we to speak of a person as knowing whatever is involved in a truth which he admits, and thus with Socrates in the *Meno* regard all reasoning as a calling to memory things already known, however forced and unnatural such a style of expression would be, this could not affect the practical importance of the course of deduction we are considering. The acorn may be said to contain the oak, but certainly will neither roof a palace, nor build a navy: still less could that implicit knowledge, which on such a view we may all be fancied to possess, supply the place of expanded wisdom. The colours of the spectrum must have been at first in the sunbeam, but are we therefore to speak of the prism as useless? Deduction is the prism of the reason; and why should we prize it more lightly, or place less faith in its power to unfold those various hues of unknown beauty, which lie veiled beneath the simplicity of truth?

But however complete the parallel between the nature of Mathematical and Moral Science, in their postulates, their definitions, their reasonings, and the certainty to which they lead, we shall be in great danger of questioning

MATHEMATICAL AND MORAL CERTAINTY. 317

the whole, unless we bear in mind the great difference which exists in the comparative ease of their attainment. The path to the latter is encompassed with many obstacles and dangers from which the former is free. Giant forms of prejudice, the idols of the market, the theatre and the cave, beset its approaches, with passion to bewilder, and pleasure to betray. It is not merely as in the other case, that the mind must be freed from distraction, and brought into the attitude of patient attention—though even this is no easy task. There is a moral atmosphere also needed, clear from the cold damps of interest, with a kindly mixture of sunshine and shower, before the germs which may have been sown in the conscience can grow up into full conviction or ripened knowledge, or root themselves firmly against the passing storm.

These observations, which are rather hints that might lead to thought, than thoughts themselves, may be found, perhaps, if pursued into their further bearings, to remove the blank which has been so often felt, and which was alluded to before. They would thus lead us to see the great moral purpose to be fulfilled, when rightly understood, by mathematical truth, as one wide emblem of the sureness of those unchanging laws, which, whether remembered or forgotten, whether honoured or despised, rule with unfailing might in the destinies of men and of kingdoms. Thus might we learn that this wand of science, which has been used with fatal force by unblest hands, to freeze the spirit into moral apathy, and bind down its powers to earth, needs only to be reversed to break the spell, and to rear it once more from the dust into the mild and stately dignity which befits the assured possessor of eternal truth. Once taught to understand this his appointed office, even the Genius of the severer sciences, blinded almost by poring continually on his figures traced in the sand, might be brought to turn his countenance heavenward, and with silent and upward eye do homage to faith's calm and moonlight beam.

A remark on this subject, in closing, may not be unsuited to the time-hallowed walls of this retreat of learning and nursery of the Christian church.

There is something in the memory of departed ages, which lends a tender and solemn feeling to every theme with which it is mingled, and flings a twilight grandeur wherever it falls. And never perhaps does the high dignity of the pursuits of science strike the soul with deeper power, than when we thus view them as a charge bequeathed by past generations, and see its signal lights thus handed onward, and still onward, over the hills of time. It is then by degrees that we are wakened into the sense of its imperishableness; and while we reflect how many, that have sought to decipher this book of nature, have long since passed away, and these mysterious characters remain unchanging and unchangeable, we gaze on it with a thrill of livelier interest, and feel that the names, which have engraven themselves on its truths, have gained a more lasting record than pillars of brass or iron. A feeling of awe thus mingles with our inquiries, as if we were searching into the archives of eternity; and if privileged to pass onward a single step in these pleasant paths, the high consciousness awakens that we are now communing with truth, which from the beginning has had its dwelling in the temple of nature, but had ever before been veiled, in still retirement, from every mortal eye.

But if such be the direct interest of natural science, when viewed as the deposit left us by the fathers of a former age, how far deeper does it become, when we learn to see in it the reflected image of truth, immutable like itself, but nobler in its objects, and loftier in its aims; a truth mingled with all the mysteries of the soul's inmost consciousness, and which the same hands have entrusted to our charge! The porch itself is worthy of all admiration —how much more the temple into which it leads! And though we may see too many who enter the former, and admire its beauty, refusing to press into the inner shrine, and even turning their backs in contempt when they ought to adore, to worship nature in her forms of eastern splendor, forgetful of Him who has robed her in the garments of light, we shall not from this feel tempted to doubt the harmony of the whole building, or the veiled glory that dwells within. The pride of an exclusive pursuit may lead the sceptic analyst to grasp all certainty as his peculiar claim, and to look with scorn upon truths too living, and too mighty, for an enfeebled conscience and a palsied heart. But those who enter into the spirit of the fathers of our church, who have here left their memory at least to their children, will read the lessons of science with a different eye. And while they train their thoughts by its severe and wholesome discipline, they will also see in it a monitor that is ever pointing them onward to holier aims. Amidst the busy scenes of life, in all their farthest and most inland wanderings, they will hear its voice, like the mystic shell from the caves of ocean, breathe in their ears its soft and solemn warning. It will thus recall their thoughts to 'that immortal sea,' towards which the spirit yearns continually, till they dwell once more in the presence of truths, unchanging and eternal, although forsaken and unvisited by the passing throng,

> "And see the children sport upon the shore,
> And hear the mighty waters rolling evermore."

www.ingramcontent.com/pod-product-compliance
Lightning Source LLC
Chambersburg PA
CBHW030735230426
43667CB00007B/723